ECOLOGICAL STEWARDSHIP

Ecological Stewardship

A Common Reference for Ecosystem Management

Volume I: Key Findings

- **Biological and Ecological Dimensions**
- **Humans as Agents of Ecological Change**
- **Public Expectations, Values, and Law**
- **Social and Cultural Dimensions**
- **Economic Dimensions**
- **Information and Data Management**

Editors

N.C. Johnson, A.J. Malk, R.C. Szaro & W.T. Sexton

A practical reference for scientists and resource managers

ELSEVIER
SCIENCE LTD

ELSEVIER SCIENCE Ltd
The Boulevard, Langford Lane
Kidlington, Oxford OX5 1GB, UK

First edition 1999

Library of Congress Cataloging in Publication Data
A catalog record from the Library of Congress has been applied for.

British Library Cataloguing in Publication Data
A catalogue record from the British Library has been applied for.

ISBN: 0-08-042816-9 (Volume I)
       0-08-043206-9 (Set: Volumes I–III)

⊚ The paper used in this publication meets the requirements of ANSI/NISO Z39.48-1992 (Permanence of Paper).

Printed in The Netherlands.

Cooperating Organisations

USDA Forest Service

USDI National Oceanic and
Atmospheric Administration

USDI Bureau of Land Management

USDI Fish & Wildlife Service

USDI Geological Survey
USDI National Biological Service

USDI National Park Service

WRI
World Resources Institute

Cooperating Foundations

American Fisheries Society
American Forests
Boise Cascade Corporation
Bullit Foundation
Consultative Group on Biological Diversity
Hispanic Association of Colleges and Universities
Liz Claiborne & Art Ortenberg Foundation
Moriah Fund
National Fish and Wildlife Foundation
National Forest Foundation
National Parks and Conservation Association
Pacific Rivers Council
Pinchot Institute for Conservation
Society of American Archeology
The Henry P. Kendall Foundation
The Nature Conservancy
The Pew Charitable Trusts
Tides Foundation
University of Arizona
W. Alton Jones Foundation

Foreword

by

Jack Ward Thomas, *Boone and Crocket Professor of Wildlife Conservation, School of Forestry, University of Montana, USA; Chief Emeritus U.S. Forest Service*

This book is a milestone on the road to a more holistic management approach to renewable natural resources. When Forest Service Chief F. Dale Robertson declared in 1992 that the agency would henceforth follow a path of "ecosystem management", I thought it was a bold move, boldly taken and at the right time. Clearly, the old multiple-use management paradigm and the associated land-use planning did not serve well when constantly impacted by the listing of one threatened or endangered species after another.

Many have been and are frightened by this new concept and considered it a revolution in land management. This was no revolution. In fact, the move to ecosystem management was simply another step — albeit a large step — in an evolutionary process. The only differences from the management approach already in place were in terms of simultaneously considering, in the course of planning and management, several new or modified factors. The areas under consideration were usually larger and more logical in terms of defining factors (ecological, economic, social, political), longer time frames are involved, additional variables are considered, and it is clearly recognized that people and their needs and desires are critical factors.

Some have argued that the term is nebulous, that ecosystems can be anything and everything, and this is but one more "buzz word". That could be true enough if the process stopped there. Ecosystem management is a concept that must always be placed in context. That context includes a clearly identified area, a prescribed time period, definition of variables that will be considered, and the detailing of the needs and desires of people and the extent to which they can be met within economic, ecological, and legal constraints. In the light of a clearly defined context, most of such charges dissipate.

Ecosystem management, in my opinion at least, is simply a concept of land management whose time has come. The concepts of the interrelationships of humans and nature are present in the religious texts underlying many religions. Some philosophers have pondered similar questions. The "subversive science" of ecology has laid the technical foundation. The now well established concern with the retention of biodiversity in the process of land management — particularly the management of the public's lands — has produced a technical management requirement to meet a political objective.

And, the combination of computers and remote sensing capability have combined with the above mentioned factors to make ecosystem management a feasible means to meet expanded land management demands.

I do not believe there is any rational retreat from this continuous and rapidly evolving concept. In a hearing before a Senate Committee on Energy and Natural Resources, I was asked where in the law it said the Forest Service could use "ecosystem management". I responded that I saw no prohibition against the approach, noted that the purpose of the Endangered Species Act was the preservation of the ecosystems upon which threatened and endangered species depended, and mentioned the judicial statement that there was no alternative that would meet the myriad requirements of law.

The Senator said he didn't understand it, didn't like it, and thought he might proceed to make "ecosystem management" against the law. I suggested that, before he did that, he might go out to the beach and practice by ordering the tide to roll back. Neither of us laughed.

It became clear to me in the course of such hassles that the concept of ecosystem management required a firm and detailed intellectual foundation to support the structure that was being put in place. This foundation would require structural stones representing many aspects of science — including natural sciences, social sciences (including economics and political sciences), and law. I felt that once this foundation was in place we could implement the concepts of ecosystem management in the next round of planning for National Forests.

I felt this process needed to be as inclusive of various technical fields and experts as possible, jointly sponsored by government and the private sector, jointly planned, and jointly executed. That was accomplished — but not without much time spent in allaying suspicion and building trust among and between participants.

The first step was the organization of this effort — including the securing of funding and commitment of the participants. The second step was an intense two-week workshop in Arizona where the various teams that put together the chapters in this book did their initial work. Each team received feedback from the overall group. When the workshop ended, the teams continued to work to produce their assigned chapters. What you have in your hands is the results of this critical step in the evolution of the concept of ecosystem management. I hope and believe that this is a foundation upon which land managers can build and rely. It is a strong foundation that will bear the weight that is placed upon it. It is what I hoped for. I trust that the authors of the chapters, the organizers of this effort, the sponsors, and the participants in the workshop will look back in a decade and say with pride, "I was there."

I, and all who contributed to this effort, owe two of my Forest Service colleagues a word of appreciation and recognition for a superb effort — Drs William Sexton and Robert Szaro. I asked much, perhaps too much, of them and they delivered even more than I asked.

Contents

Foreword vii
Preface xi
Acknowledgments xvii

Introduction—The Ecological Stewardship Project: A Vision for Sustainable Resource Management . . 1

BIOLOGICAL AND ECOLOGICAL DIMENSIONS
Genetic and Species Diversity 13
Population Viability Analysis 19
Ecosystem and Landscape Diversity 27
Ecosystem Processes and Functioning 33
Disturbance and Temporal Dynamics 39
Scale Phenomena 47
Ecological Classification. 53

HUMANS AS AGENTS OF ECOLOGICAL CHANGE
Human Roles in the Evolution of North American Ecosystems 63
Cultural Heritage Management 71
Producing and Using Natural Resources 77
Ecosystem Sustainability and Condition 85
Ecological Restoration. 95

PUBLIC EXPECTATIONS, VALUES, AND LAW
Legal Perspectives 107
Public Expectations and Shifting Values 115
The Evolution of Public Agency Beliefs and Behavior. 123
Processes For Collaboration. 129
Regional Cooperation 135

SOCIAL AND CULTURAL DIMENSIONS
Cultural/Social Diversity and Resource Use 145
Social Classification 151
Social Processes. 159

ECONOMIC DIMENSIONS
Demographics and Shifting Land and Resource Use 169
Economic Interactions at Local, Regional, and National Scales. 177
Ecological and Resource Economics 185
Uncertainty and Risk Assessment 191
Economic Tools For Ecological Stewardship 197

INFORMATION AND DATA MANAGEMENT

Adaptive Management 207
Assessment Methods. 215
Monitoring and Evaluation 223
Data Collection, Management, and Inventory 231
Decision Support Systems. 241

Index 247
List of Abbreviations 277
Contents of Volumes I–III 281

Preface

by

William T. Sexton, *Deputy Director, Ecosystem Management, U.S. Forest Service*
Robert C. Szaro, *Special Assistant, Ecosystem Management Research, U.S. Forest Service*
Nels C. Johnson, *Deputy Director, Biological Resources Program, World Resources Institute*
Andrew J. Malk, *Associate, Biological Resources Program, World Resources Institute*

> *"The difficulty lies, not so much in developing new ideas,*
> *as in escaping from the old ones"*
> — John Maynard Keynes

For two weeks in December of 1995, over 350 natural resource managers, ecologists, economists, sociologists, and administrators gathered in Tucson, Arizona, to begin developing a knowledge base for the stewardship of lands and waters in the United States. As Jack Ward Thomas, then Chief of the U.S. Forest Service, noted at the opening of the workshop, better access to scientific findings and management experiences is a key to promoting wider use of ecosystem approaches in natural resource management. The teams of resource managers and researchers assembled in Tucson were invited on the basis of their knowledge and experience on issues central to implementing ecosystem approaches to natural resources management. Over the next two years, these teams labored to synthesize both current scientific understanding and resource management experiences on ecological, social, economic, and resource management topics. This reference is the final product of their efforts. It is intended to be useful to resource managers in the field, planners and decision makers, and researchers seeking to better understand ecosystems and the people who interact with them.

The contributors, the management examples, and the audience of this reference are largely located in the United States. Nevertheless, as organizations around the world increasingly turn to ecosystem management, bioregional management, and other integrated approaches to natural resource management, we hope they will find the process and information generated by the Ecological Stewardship Project useful in their own context.

DISTINCTIVE FEATURES OF THE ECOLOGICAL STEWARDSHIP REFERENCE

The process used to develop the *Ecological Stewardship* reference was distinctive in several respects. First, recognizing that no one organization could marshall the understanding and experience to comprehensively address ecosystem approaches, partnerships became an essential feature of the Ecological Stewardship project. Second, recognizing that implementation depends on good science and lessons from the field, the project strived to link both research and management perspectives on the topics it addressed. Third, recognizing that accuracy is a key feature of a reference work, the project used a peer-reviewed forum to encourage discussion and documentation of competing views, philosophies, and facts. Fourth, recognizing that the reference needs to be useful in virtually any natural resource management setting, the project sought diverse examples and avoided prescriptive recommendations and site-specific goals. Fifth, recognizing the need to make the reference widely accessible, the project has used the Internet both to facilitate comments on draft chapters and to display the final document. Finally, recognizing the large volume of information contained in the reference and the limited time of resource managers and other users, the project has developed a concise key findings volume and a CD-ROM version of the full text.

In particular, users of this reference should understand the key role partnerships played in developing this document, and the importance of the review process.

Partnerships

This reference exists only because a large number of government agencies, private foundations, interest groups, and individuals committed themselves to share their knowledge, experiences, time, and financial resources. Two types of partnerships were especially important. First, a partnership between public agencies and private organizations helped to ensure the broad usefulness of the reference. Although managers of federal lands and waters are the primary audience, we are confident the information it contains will also be useful to managers of state and private lands, and to researchers and analysts in universities and non-governmental organizations. Second, a partnership between researchers and resource managers helps to ensure that the reference is both scientifically sound and addresses the issues most often encountered by natural resource management practitioners.

A Public–Private Partnership

The need to develop partnerships between public and private organizations was driven by several factors. First, neither the U.S. Forest Service nor any other federal agency had all of the experience or experts needed to synthesize state-of-the-art knowledge about ecosystem approaches. Second, to be most useful, the reference needed to be relevant to a range of organizations involved in natural resources management. A partnership between public agencies and between public and private organizations has helped to ensure broad ownership of the document. Third, the significant cost in human and financial resources to develop the reference would be difficult for any one organization to bear. Financial commitments by private foundations and commitments of human resources by various federal agencies were vital to making the Ecological Stewardship Project work.

The use of partnerships characterized the project from the beginning (Box 1). Dozens of individuals from public agencies and private organizations

---

Box 1. Chronology of Major Steps in the Ecological Stewardship Project

April–September 1994:	Project concept developed by Jack Ward Thomas and the eight foundations of the Consultative Group on Biodiversity.
October 1994–August 1995:	Public meetings to identify topics and workshop participants.
December 1995:	Ecological Stewardship Workshop, Tucson, Arizona.
May 1996:	Cooperative Agreement signed between U.S. Forest Service and World Resources Institute to develop Ecological Stewardship reference.
June 1996	Rough draft chapters reviewed by editorial team.
January–April 1997:	Expert and public review of draft chapters.
April–May 1997:	Review panel meetings in Portland, Oregon.
December 1997:	Final chapter drafts for Volumes II and III submitted to World Resources Institute.
May 1998:	Publication agreement developed with Elsevier Science Ltd.
April–August 1998:	Preparation of Key Findings volume.
September 1999:	Publication of Ecological Stewardship Reference by Elsevier Science Ltd.

---

met regularly in the first months of the effort to identify topics to be addressed, and to identify individuals who could contribute as authors and participants in the Tucson workshop. Eight private foundations and several companies made generous financial contributions early on. The workshop itself involved participants from federal and state agencies, universities, private resource management companies, and non-governmental organizations. Among the authors are staff from six federal agencies, several state agencies, over a dozen universities, several forest products companies, and several non-governmental organizations. To ensure the credibility of the effort, the U.S. Forest Service engaged the World Resources Institute to develop an independent review process and oversee development of the reference materials. Finally, to enhance marketing and accessibility, Elsevier Science Ltd was engaged to publish the reference.

The broad partnership used to develop this reference is representative of the cooperation essential to any sound ecosystem approach to natural resources management. Not surprisingly, examples of collaborative partnerships are found throughout the document. Many chapters stress both technical and practical reasons for using partnerships in ecosystem approaches to natural resource management. Developing a knowledge base, like other elements of an ecological approach, requires close collaboration between many resource management interests and organizations. During its four years of development, the Ecological Stewardship Project engaged literally hundreds of individuals and dozens of organizations.

A Partnership of Researchers and Managers

Communication between scientists and resource managers is a fundamental requirement of ecosystem approaches to natural resources management. A sound scientific foundation is needed to develop credible management options that both work and can withstand legal challenges. Managers, like scientists, are concerned with the validity of their information. A common thread for decision making, therefore, is that all stakeholders have access to the most scientifically sound information available. At the same time, some of the most important challenges facing resource managers are related to social, economic, political and institutional factors. Research can shed important light on these factors, but implementation experiences are needed to provide insights into what works and what does not.

From the start, the Ecological Stewardship Project has sought to foster a partnership between researchers and managers. Teams of researchers and managers were assembled to examine both the science and the management experiences behind the topics addressed in this reference. At the Tucson workshop, scientists and managers critiqued each others' outlines before writing began. When review drafts were completed, steps were taken to ensure that each chapter was reviewed by both researchers and managers. Staff from the World Resources Institute visited with natural resource managers in Montana, Maryland, and West Virginia to seek their views on how the information could be made most useful to them. The format and content of the Key Findings volume (Volume I) are a direct result of those conversations. Ecosystem approaches afford new opportunities to strengthen partnerships between researchers and managers in ways that will benefit both science and management. It is our hope that this reference will contribute to that process.

A Rigorous and Public Review Process

Organizers of the Ecological Stewardship Project viewed an independent and rigorous review process as essential to the project's credibility. To achieve this, the U.S. Forest Service and the World Resources Institute (WRI) entered into a cooperative agreement in which WRI would design and manage an independent peer review process and oversee the development of the manuscript.

In consultation with the U.S. Forest Service and a steering group composed of representatives of federal and state agencies, non-governmental organizations, and industry, WRI designed a three-stage peer review process. The first review stage, completed in August 1996, provided authors with general comments from the editorial team. The main goal was to strengthen the links between research and management chapters on the same topic, and identify revisions needed before chapters were ready for a more formal review.

A formal review process for the revised chapters took place between January and April of 1997. Each chapter was reviewed by at least three, and typically more, recognized experts including both researchers and managers. At the same time, WRI created a World Wide Web site to facilitate public access to the draft papers. Notices were placed in a number of professional journals (e.g., *Conservation Biology, Society and Natural Resources, Journal of Forestry, Journal of Range Management,* and *Journal of Ecological Economics*) and on computer list-serves to invite broader participation in the review process.

To facilitate the review, WRI divided the chapters into six thematic groups, each with a coordinator to summarize reviewer comments and help authors to identify the most important revisions. The groups and their coordinators included: *Public Values and Expectations* (Dean Bibles, Department of Interior); *Social and Cultural Dimensions* (Margaret Shannon, University of Syracuse); *Humans as Agents of Ecological Change* (Nels Johnson, World Resources Institute); *Biological and Ecological Dimensions* (Gary Meffe, University of Florida);

Economic Dimensions (Robert Mendelsohn, Yale University), and *Data and Information Management Tools* (Francisco Dallmeier, Smithsonian Institution).

After the reviews had been received and summarized by the coordinators, a "review panel" was convened for each thematic group. Chaired by the thematic group coordinator, these panels brought together lead authors from each chapter in the group and outside experts. The panel meetings accomplished two objectives. First, they enabled authors to meet with their coordinator to agree on how to revise their final drafts. Second, they enabled the lead authors and outside experts assigned to their topic to summarize the key findings from their chapters. These summaries provided the "raw" material used by the editorial team to draft the Key Findings volume.

ORGANIZATION OF THIS REFERENCE

The complete *Ecological Stewardship* work consists of a Key Findings volume (Volume I) and two volumes containing 54 chapters that address 31 separate topics (Volumes II & III). The original plan was to have a total of 30 topics, each addressed by two chapters — one on the scientific foundations of the topic, the other on management perspectives and experiences. This symmetry did not survive for several reasons. For several topics, the science and the management author teams decided to integrate their chapters. In several other cases, the

---

Box 2. Topics Addressed by the "Ecological Stewardship" Reference

The 31 topics addressed in the reference are grouped in six thematic sections

Volume II

BIOLOGICAL AND ECOLOGICAL DIMENSIONS
- Genetic and species diversity
- Population viability
- Ecosystem/landscape diversity
- Ecological functions and processes
- Role of disturbance and temporal dynamics
- Scale phenomena
- Ecological classification

HUMANS AS AGENTS OF ECOLOGICAL CHANGE
- Human role in evolution of North American ecosystems
- Cultural heritage management
- Ecosystem sustainability and condition
- Ecological restoration
- Producing and using natural resources

Volume III

PUBLIC EXPECTATIONS, VALUES, AND LAW
- Public expectations and shifting values
- Processes for achieving consensus
- Regional cooperation
- Evolving public agency beliefs and behaviors
- Legal perspectives

SOCIAL AND CULTURAL DIMENSIONS
- Cultural/social diversity and resource use
- Cultural heritage management
- Social/cultural classification
- Social system functions and processes

ECONOMIC DIMENSIONS
- Shifting demands for natural resources
- Economic interactions at local/regional/national scales
- Ecological economics
- Uncertainty and risk assessment
- Economic tools and incentives for ecological stewardship

INFORMATION AND DATA MANAGEMENT
- Adaptive management
- Monitoring and evaluation
- Data collection, management, and inventory
- Assessment methods
- Decision support systems/models and analysis

author team was either unable to finish their chapter or was unable to make satisfactory revisions following the review process. For topics where one chapter did not make it to completion, efforts were made to integrate at least some of the missing scientific or management perspective into the remaining chapter. Finally, one topic was added to address a large gap identified by the editorial team.

The topics addressed by the *Ecological Stewardship* are listed in Box 2. For a complete listing of all chapters, please turn to the complete table of contents listed in the back of each volume.

Key Findings Volume (Volume I)

The Key Findings volume is designed to highlight both scientific findings and management insights for all 31 topics addressed by the *Ecological Stewardship* reference. The topic summaries are not executive summaries of individual chapters nor are they referenced. Rather, they highlight key findings and practical considerations that come from the one or two chapters that address a particular topic. Rough drafts for most of these summaries were prepared by teams assigned to each topic at the panel review meetings in Portland, Oregon. Final drafts were prepared by the editorial team once the chapters had been completed. Wherever possible, the editorial team sought to use language directly from the individual chapters. Otherwise, key findings and practical considerations were distilled from the full content of the relevant chapter(s). To ensure that the topic summaries are faithful to the contents of the chapters on which they are based, the final drafts were sent to each of the lead authors for their review.

We think the Key Findings volume is an extremely useful tool that enables readers to quickly learn about the full range of topics addressed by the *Ecological Stewardship* reference. To learn more about a given topic, we urge readers to turn to the relevant chapters in Volumes II and III.

Volumes II and III

Volumes II and III contain 54 chapters prepared for the *Ecological Stewardship* reference. Volume II begins with three chapters that provide broad cross-cutting analysis and perspectives on ecosystem approaches to natural resources management. The rest of the volume contains the thematic sections on Biological and Ecological Dimensions (13 chapters) and Humans as Agents of Ecological Change (9 chapters). Volume III contains thematic sections on Public Expectations, Values and Law (7 chapters), Social and Cultural Dimensions (5 chapters), Economic Dimensions (8 chapters) and Information and Data Management (9 chapters). Each chapter is fully referenced, and cross references have been made by the editorial team to other chapters that address issues in greater detail. Each thematic section begins with an 8–10 page overview by the coordinator for that group of papers. The purpose of the thematic overview is to give readers a sense of the linkages between the chapters in the section. We urge users to read the overviews since they place individual chapters in a larger context.

CD-ROM

Finally, the *Ecological Stewardship* reference is accompanied by a CD-ROM containing the complete text of all three volumes. This CD-ROM is intended to serve as a tool to help users locate specific information, and is searchable by table of contents, keywords and index.

Acknowledgments

Partnerships have been fundamental to the Ecological Stewardship Project from the initial concept meeting that launched the project in 1994 to the publication of this reference work nearly five years later. Several hundred people have been involved in some facet of the project during that time as planners, authors, reviewers, and workshop organizers. Each of those individuals contributed support, energy, and ideas that helped make this project a success. A number of individuals and organizations, however, merit special recognition for their efforts.

Jack Ward Thomas, Chief of the Forest Service from 1993 to 1996, defined the need for a reference work that addressed both the scientific fundamentals of ecosystem management and lessons and perspectives from resource managers in the field. His vision of the project included a broad-based partnership of public agencies, private foundations, and researchers and managers from state and federal government, universities, non-governmental organizations, and the private sector. Without his energy and commitment, this reference would not exist.

Some of the earliest and most important partners in the Ecological Stewardship Project have been several private foundations including the Bullit Foundation, Henry P. Kendall Foundation, Liz Claiborne & Art Ortenberg Foundation, Moriah Fund, National Fish and Wildlife Foundation, Pew Charitable Trusts, Tides Foundation, and the W. Alton Jones Foundation. In addition to providing vital financial support, they also supplied ideas and guidance during the early conceptualization of the project. Peter Stangel of the National Fish and Wildlife Foundation was instrumental in coordinating support from the foundations and as an advisor and contributor throughout the project's evolution. Financial contributions were also provided by the Hancock Timber Resources Group and the Seneca Jones Timber Company.

Several federal agencies co-sponsored the project including the Bureau of Land Management, U.S. Fish and Wildlife Service, U.S. Geological Survey, U.S. National Park Service, U.S. Forest Service, and the National Oceanic and Atmospheric Administration. Beyond co-sponsorship, these agencies enabled many of their talented staff to actively contribute to the project as workshop participants, authors, and reviewers. In addition, a number of professional societies, conservation groups, and corporations co-sponsored the Tucson Workshop that brought together over 300 participants to plan and outline the contents of this reference. This included the American Fisheries Society, American Forests, Boise Cascade Corporation, Consultative Group on Biodiversity, Hispanic Association of Colleges and Universities, National

Forest Foundation, National Parks and Conservation Association, Pacific Rivers Council, Pinchot Institute for Conservation, Society of American Archeology, The Nature Conservancy, and the University of Arizona. Individuals from these and other organizations participated in a five month series of public meetings held in Washington to plan and shape the project in the summer and fall of 1995.

The December 1995 workshop in Tucson, Arizona brought together all of the authors and a wide range of scientists and resource managers to develop a framework for the topics and papers that make up this reference. Special thanks go to those on the workshop organizing team, particularly Pam Kelty, to the facilitators from various agencies who kept things on track for two intensive weeks, and to the personnel who provided indispensable logistical and administrative support during the workshop. Larry Hamilton, BLM State Director in Montana was exceptional as Master of Ceremonies and guiding light for the workshop; his efforts are greatly appreciated. A small group who helped identify key cross-cutting themes and follow-up issues at the Tucson Workshop included Rai Behnert, Gary Benson, Gordon Brown, Jim Caplan, Bill Civish, Jerry Clark, Joan Comanor, Bob Doppelt, Chris Jauhola, Dennis LeMaster, John Mosesso, Steve Ragone, Chris Risbrudt, Bob Robinson, Peter Stangel, Brad Smith and Dick Smythe. Finally, the University of Arizona served as host and co-sponsor for the workshop. Pat Reid and Carol Wakely at the University were critical to making the Tucson Workshop a success.

A steering group helped the World Resources Institute define the review process and explore publication options. Members of that group included Greg Aplet, Douglas Burger, Robert Dumke, Judy Guse-Noritake, John Haufler, Meg Jensen, John Mosesso, Jeff Olson, Chris Risbrudt, Walt Reid, Larry Ross, Craig Shafer, Margaret Shannon, Steve Thompson, and Henry Whittemore. Their suggestions helped to shape the reference work in several important ways, most notably establishing the importance of having a relatively brief Key Findings volume. Six "thematic group coordinators" — Dean Bibles, Francisco Dallmeier (assisted by Deanna Kloepffer), Nels Johnson, Gary Meffe, Rob Mendelsohn, and Margaret Shannon — were indispensable to the review process and helping authors revise their drafts. We are also grateful to Clint Carlson, Butch Farabee, Steve Kelly, Kniffy Hamilton, Larry Hamilton, Bertie Pearson, and Hal Salwasser in Montana, and to Julie Concannon in West Virginia and to Mark Koenig and Carl Zimmerman in Maryland for their help in arranging interviews with field resource managers. These interviews helped us to design this reference, particularly Volume I.

The reviewers (see list below) are of particular note. They not only reviewed lengthy manuscripts with no compensation, but many of them actually reviewed two manuscripts (one on the research side and one on the management side of the same topic). Their reviews provided the grist for a series of review panel meetings in Portland, Oregon during the spring of 1997. Panel participants (see list below) helped authors focus on key conclusions and findings for their chapters and helped draft the raw materials for the Key Findings volume. Thanks go in particular to Trish White, Oretta Tarkhani, Claudia Tejani, and Teri Aledo who helped organize the intensive three week series of review panel meetings like clockwork.

Robert Hamre and David Johnston edited final draft papers. Their skills added immeasurably to the readability and consistency of the individual chapters in this reference. Carolyne Hutter edited the Key Findings volume and the introductory overviews for the thematic sections. Their skill under tight deadlines, as well as their patience and good humor, are deserving of special recognition here. Trish White stepped in with much needed background research and assistance to put the Key Findings volume together at the very end.

Finally, we are grateful for Mary Malin's enthusiasm and guidance at Elsevier Science. She and others at Elsevier, including Pam Birtles, transformed an enormous amount of manuscript into an attractive and finished product.

PROJECT SPONSORS

Many private foundations, corporations, organizations and federal agencies played a critical role in making this vision a reality by accelerating the collaborative process and setting the stage for the innovative public–private partnership necessary to develop this reference document. Public–private partnerships are at the core of an ecological approach and this project "lived the vision" as a major element in making this reference possible.

Private Financial Supporters

Bullitt Foundation
Hancock Timber Resources Group
Liz Claiborne & Art Ortenberg Foundation
Moriah Fund
National Fish and Wildlife Foundation
Seneca Jones Timber Company
The Henry P. Kendall Foundation
The Pew Charitable Trusts
Tides Foundation
W. Alton Jones Foundation

Agencies and Organizational Supporters

American Fisheries Society
American Forests
Boise Cascade Corporation
Consultative Group on Biological Diversity
Hispanic Association of Colleges and Universities
National Forest Foundation
National Parks and Conservation Association
Pacific Rivers Council
Pinchot Institute for Conservation
Society of American Archeology
The Nature Conservancy
University of Arizona
USDA Forest Service
USDI National Oceanic & Atmospheric Admin
USDI Bureau of Land Management
USDI Fish and Wildlife Service
USDI Geologic Survey
USDI National Biological Service
USDI National Park Service
World Resources Institute

CHAPTER REVIEWERS AND PANEL PARTICIPANTS

The following individuals generously gave their time to review chapters and/or participate in review panel meetings that were held in Portland, Oregon in April 1997.

James Agee, Paul Alaback, Michael Alvard, Steve Apflebaum, Greg Aplet, Steve Arno, Richard Artley, Jerry Asher, Gordon Bakersville, David Bayles, Jill Belsky, Dean Beyer, Dean Bibles, William Birch, Jr., Myron Blank, John Bliss, Bernard Bormann, Dale Bosworth, Bill Bradley, Martha Brookes, Deborah Brosnan, Gordon Brown, Michael Brown, Perry Brown, Donald Callaway, Henry Campa, David Capen, David Caraher, Clint Carlsen, Connie Carpenter, C. Ronald Carroll, Joyce Casey, Alfonso Peter Castro, Jiquan Chin, Steve Cinnamon, Roger Clark, Beverly Collins, Julie Concannon, Allen Cooperrider, Hanna Cortner, W. Wallace Covington, Paul Cunningham, Francisco Dallmeier, Terry Daniel, Steve Daniels, Roy Darwin, Frank Davis, Robert Dumke, David Edelson, Robert Ewing, Stephen Farber, Scott Farrow, Scott Ferson, Suzanne Fish, Cornelia Flora, Jo Ellen Force, Tim Foresman, Sandra Jo Forney, Lee Frelich, Oz Garten, Paul Geissler, Susan Giannentino, James Gosz, Russell Graham, Dennis Grossman, Elaine Hallmark, Johnathan Harrod, Johathan Haufler, Richard Haynes, Robert Hedricks, Gerald Helton, Thomas Hemmerly, Al Hendricks, Karen Holl, Tom Holmes, Anne Hoover, Amy Horne, Mark Hummell, Michael Huston, Donald Imm, Patrice Janiga, David Jaynes, Robert Jenkins, Mark Jensen, Meg Jensen, Kathleen Johnson, David Johnston, Ron Kaufman, Robert Keiter, Jim Kenna, Winnie Kessler, Kate Kitchell, Michael Klise, Deanne Kloepfer, Dennis Knight, Richard Krannick, Linda Langner, Bill Lauenroth, Daniel Leavell, Susan Lees, Gene Lessard, Bernard Lewis, Henry Lewis, Joseph Lint, Bruce Lippke, Doug Loh, Mark Lorenzo, B. John Losensky, Ariel Lugo, Douglas Mac Cleery, David Maddox, Jerry Magee, Marie Magleby, Robert Mandelsohn, Jon Martin, Clay Mathers, Mary Mc Burney, Maureen Mc Donough, Joe McGlincy, Gary Meffe, Gene Namkoong, Judy Nelson, Barry Noon, James O' Connell, Joseph O' Leary, Carlton Owen, J. Kathy Parker, Steve Paustain, Brian Payne, John Peine, Richard Periman, David Perry, Chris Peterson, Dave Peterson, George Peterson, Steward Pickett, Tom Quigley, Carol Raish, Connie Reid , Carl Reidel, Fritz Rennebaum, Keith Reynolds, Thomas Riley, Jeffrey Romm, William Romme, Len Ruggiero, Michael Ruggiero, Linda Rundell, Fred Samson, Don Schwandt, Jim Sedell, Roger Sedjo, Jan Sendzimir, William Sexton, Mark Shaffer, Margaret Shannon, William Shaw, Thomas Sisk, William Snape, Joel Snodgrass, Rhey Soloman, Jim Space, Pat Spoerl, Cleve Steward, Victoria Sturtevant, Alan Sullivan, Robert Szaro, Joseph Tainter, Paul Templet, Timothy Tolle, David Wear, Patricia White, Henry Whittemore, David Williams, Sue Willits,

Introduction

The Ecological Stewardship Project:
A Vision for Sustainable Resource Management

by

William T. Sexton, *Deputy Director, Ecosystem Management, U.S. Forest Service*
(Co-executive Secretary of the Ecological Stewardship Project)
Robert C. Szaro, *Special Assistant, Ecosystem Management Research, U.S. Forest Service* (Co-executive Secretary of the Ecological Stewardship Project)
Nels C. Johnson, *Deputy Director, Biological Resources Program, World Resources Institute* (WRI Ecological Stewardship Project Manager)

Just as sustained-yield and multiple-use concepts emerged as frameworks to manage natural resources earlier in this century, the end of the 20th century has become an era of change in the use of knowledge, tools, and strategies to manage lands and waters in the United States. The shift now is to ecological stewardship approaches that enable public and private resource managers to use science plus social guidance to help sustain productivity while maintaining biodiversity and other ecosystem services. This conceptual shift is evolving rapidly, as measured by the emergence of new concepts of resource management, an expanding body of scientific information relevant to management practices, a growing number of ecosystem management efforts around the country, and the development of new policies (Cortner and Moote 1998).

Yet, the concept and practice of integrating ecological knowledge into natural resources management — often referred to as "ecosystem management" — is not altogether new in the United States. For example, Aldo Leopold (Leopold 1949) was advocating the use of ecological knowledge in natural resources management as early as the 1940s and by the 1970s ecological concepts were finding their way into a growing number of publications on wildlife and natural resources management (e.g., Thomas 1979). More recently, the concept of "sustainable development" — often defined as managing natural resources to meet present human needs without compromising the ability of future generations to meet their needs — has led to greater consideration of the long term impacts of natural resource management decisions. Meanwhile, the cumulative impacts of numerous local management decisions have led many scientists and resource managers to conclude that biodiversity, water quality, and other natural resources can only be conserved through cooperative efforts across large landscapes — landscapes that often cross ownership boundaries (e.g., Aplet et al. 1993; PCSD 1996; Szaro and Sexton 1996). Researchers and resource managers in other parts of the world have also recognized the value of integrated regional approaches to conservation and resource management, although terms such as "bioregional management" and "integrated conservation and development projects" may be used in place of "ecosystem management" (Miller 1996).

What distinguishes ecosystem management approaches from earlier strategies and approaches to natural resources management is the integration of environmental, social, and economic knowledge to make sustainable resource management decisions at multiple geographic scales (Sexton et al. 1996). Shifting from single resource or single species management to managing an ecosystem for a variety of resources, including its biodiversity, requires using the best available scientific, social, and economic information (PCSD 1996). Scientific information is needed to identify ecosystem processes essential to the productivity of a wide variety of natural resources. Social and economic information can be used to determine which strategies will best meet public demands and landowner objectives. Voluntary and cooperative efforts at managing natural resources across ownership boundaries characterize successful ecosystem management efforts (Keystone Center 1996).

Ecosystem management, as an approach to understanding and managing the environment, provides resource managers with a broad range of tools to establish partnerships, collect information, analyze management options, and monitor the results of management and policy decisions. It includes understanding resource management options through interpreting the range of social, cultural, and spiritual values, perceptions and goals held by people, as well as the history, status and possible future trends of the physical and biological elements of the landscapes of particular interest. Using an ecosystem approach is not a simple, static prescription for a quick and perfect solution. An ecosystem approach, however, does provide a framework and a much larger tool kit for incorporating the best available science and management experience in a dynamic process of framing and addressing questions about the environment and the human presence and activity that affects it (Sexton 1998). It implies a heuristic process to address how humans think about, understand and analyze options for managing this relationship in sustainable ways. It requires that natural resource managers collect, organize and plan, and make decisions with an extensive array of scientific information and management experience.

Still, the use of ecosystem approaches faces a number of challenges. Most efforts are relatively recent, making it difficult to assess the results of ecosystem management approaches in a range of field situations (Yaffee et al. 1996). For example, nearly all the examples surveyed by the Keystone Dialogue (Keystone Center 1996) started after 1990. Another problem is that the concepts, science, and policies of ecosystem management have outpaced the rate at which managers can develop and implement practical applications as a part of their routine management practices. Much of the rapidly growing scientific literature on the subject is presented in idealized form, only limited quantitative assessments of on-the-ground experiences have been conducted, and ecosystem management concepts are continuing to evolve as field experience provides a means to understand those concepts in a context of real-world implementation. This leaves natural resource managers with limited guidance on how to apply ecosystem management concepts in specific field situations.

The use of ecosystem approaches in day-to-day management, research, and policy development lags behind, in part for want of practical assessment and a summary of the growing scientific, social, and management experience that already exists. Natural resource managers across the country are faced with the daily task of finding effective ways to manage resources sustainably in the context of a wide variety of desired conditions. Ecosystem approaches require more information about larger areas. How then do managers, often stationed in remote field locations, learn about, find, collect, organize and use information across an expanded range of disciplines?

Within the last several years, thousands of publications have addressed issues important to ecosystem approaches to natural resources management such as: the evolution of ecological concepts, perspectives on the results of past resource management practices, biodiversity inventory and conservation, a renewed emphasis on the concept of sustainability and an on-going debate about what is the appropriate interaction of humans with their environment. The result is an enormous pool of ideas, information, and critiques regarding resource management. For any individual, accessing and synthesizing this information and knowledge is difficult, if not impossible. This reference seeks to fill that gap by synthesizing both existing scientific understanding and management experience on a wide range of topics relevant to ecosystem approaches in natural resources management. It provides information, advice, concepts, and experience from many individuals in similar operating situations to aid field managers in their demanding daily routines and to help researchers define future scientific investigations.

KEY THREADS THAT LINK ECOSYSTEM MANAGEMENT APPROACHES

There is no one universal formula for defining and implementing an ecosystem approach in natural resources management. Discussions regarding the definition of what constitutes an ecosystem approach to natural resource management invariably focus on certain common key themes and linkages (e.g. Interagency Task Force 1995; Keystone Center 1996). During the Tucson workshop, a subset of the participants were asked to step back from the detailed discussions on individual topics and identify the behaviors, attitudes, and values that support a successful ecosystem approach to natural resources management.* The following points are a synthesis of the issues raised by that group.

Keeping Everyone "In the Loop"

It is critically important to involve key parties early in the ecosystem management planning process, and keep them productively involved throughout. This seems trite, but in case after case, successful planning efforts involved key parties, and unsuccessful efforts failed to do so. Agency decision-makers should consider involving the public in non-traditional ways, such as in data collection and monitoring. Managers also need to do more than just provide a

*Rai Bennett, Gary Benson, Gordon Brown, Jim Caplan, Bill Civish, Jerry Clark, Joan Comanor, Bob Doppelt, Chris Jauhola, Dennis LeMaster, John Mosesso, Steve Ragone, Chris Risbrudt, Rob Robinson, Peter Stangel, Brad Smith, and Dick Smythe.

forum for input; if necessary, they have to draw others into the process. Often, long-term residents have in-depth knowledge of how ecosystems have responded to past influences, and can provide valuable guidance not available elsewhere. Decision-makers must be receptive to such input from the public, and keep the decision-making process open. For some public interest groups, the *process* used in inviting participation is as important as the outcome.

Sharing Information for Success

To keep key parties actively engaged in the ecological stewardship process, agency personnel must share information about the ecosystem, appropriate laws, agency management goals, the decision-making process, different stakeholder interests, and a host of other factors. It may also be necessary for key players to educate agency personnel about local cultures and historical traditions. This give-and-take of information contributes to each partner's education, and builds the bonds necessary for effective ecological stewardship. A shared literacy and awareness about ecosystems, and about how humans rely on and affect them, can help build receptivity and support for ecological stewardship approaches.

Fostering a Culture of Ecological Stewardship

Social commitment exists for environmental protection, but social support for ecological stewardship can be improved. Society must be convinced that people are inextricably wedded to ecosystems and, therefore, humans have to take into account how they "draw" upon ecosystems for their personal well-being. An imperfect but useful analogy is how people routinely review and balance their checkbooks; they draw upon their resources to support their lifestyle choices but cannot withdraw beyond a minimum balance. We must also increase acceptance of inherent ecosystem value — the belief that ecosystems have intrinsic value as functioning systems as well as for how their components can be used by people. In addition, this belief asserts that system components and their relationships constitute a valued and valuable asset for all species, especially humans. Ecological stewardship is viewed as a way to bring communities together and improve the quality of natural resources. It is about looking for collaborative approaches among all landowners who desire health and productivity for the lands, waters, and resources they manage.

Motivated by a Love of Place

Successful approaches to ecosystem management are increasingly community-based — initiated by local people, and motivated by a "love of place." In such cases, ecosystem management is in the local interest, and is a means to achieve the beneficial use of its "natural capital." There is growing realization that a long-term approach to land use and management is generally better ecologically and economically for developing harmonious and sustainable relationships between people and the land. Ecosystem approaches are also a means to build trusting relationships, often among former antagonists, to gain political power in furthering common envisioned ends.

Interagency Collaboration

Interagency collaboration is essential, and can be improved, if people focus on each agency's relevant legislative authorities as *enabling* cooperation, rather than hindering it. Positive relationships between employees in the various agencies are important to success, and these should be recognized and encouraged through administrative channels. Having a shared vision for ecological stewardship is also critical for interagency collaboration. The history of agency missions has led to a somewhat disjointed federal conservation mandate, but ground-breaking efforts such as the Ecological Stewardship project bring executive leadership teams in direct contact to discuss shared visions and objectives. Many saw the networking at staff and executive levels as a productive means to continue development of a shared vision for ecological stewardship. Effective collaboration could also help overcome budgetary restrictions. Congress and other funding entities should respond more favorably to clearly demonstrated collaborative approaches to land management.

Private Industry Leadership and Partnership

Partnerships that cross-cut public/private boundaries are also essential for ecological stewardship. Awareness of private property rights was a high priority, and the notion of federal or state dominion over private lands was flatly rejected. In fact, many participants noted that private industry was providing effective models that can strengthen federal land management, and industry was invited to assume the leadership role on some resource issues.

Effective collaboration between the public and private sectors will require a shared vision and strategy for implementation, mutually acceptable and binding rules for collaboration, and willingness to trust and act separately and together to attain this vision. Shared leadership will be critical to collaboration. It recognizes "co-dominance" among partners based on strengths, talents, and expertise. It recognizes the potential for "co-evolution" in personal and institutional growth and agreements. Thus, shared leadership is really "co-leadership." Co-leadership means conscious avoidance of "I win; you lose" approaches. Co-leadership emerges when partners find common vision, establish agreements for how to treat one another, and contribute the resources for joint activity. In the case of federal agencies, laws (e.g., National Environmental Policy Act, Federal Land Policy and Management Act, National Forest Management Act, and Administrative Procedures Act) compel certain minimum actions, and these should be used in a way that maximizes tangible and effective participation by other co-leaders.

Hierarchical Approaches

Hierarchies are useful in ecological, economic, social, temporal, and political arenas to match the question or issue to the correct scale for analysis or decision-making. For example, in designing projects, knowledge of the characteristics and probable responses of the ecosystem are key to achieving objectives. It is useful to look one level above, to gain knowledge of context, and one level below, to gain understanding of content and processes. Theoretically, a hierarchy in the decision-making process for ecosystem management could include the individual, family, community, county, state, region, national, and global levels.

Agencies also have administrative hierarchies. For example, the Forest Service has ranger districts, forests, regions, the Washington Office, and the Department of Agriculture. Policies and decisions must be understood in the context of the hierarchy in which they are made.

Cultural hierarchies are sometimes overlooked, but are also critically important. Local groups may differ more in their willingness to take risks than will state or national groups. For example, a local development project may pose risks to an endangered species. Local groups may support the project, accepting the risk in favor of economic or other returns. National groups charged with the range-wide stewardship of the endangered species may not be willing to accept that same risk, given the potential impact to the species as a whole. In situations such as this, hierarchical analysis can more clearly portray the effect of decisions at multiple scales, and thereby assist with decision-making.

Shifting Cultural Values

Values shift over time — groups should not be seen as representing the same point of view on resource issues forever. Also, the public trust is constantly being re-evaluated, and reliance on older stereotypes and role models may greatly delay the implementation of a more holistic approach to resource management. Agencies should work to improve and share data on human demographic trends so that local managers can better track and predict changes that are likely to occur as a result of shifting cultural values. And, agencies must meet rapid cultural changes with rapid and effective personnel changes and training.

The Role of Science in Ecosystem Management

Everyone understands that good science is critical to good decision-making, but much research is not relevant to the actual decisions that managers must make. In particular, there is increasing need for research that looks at structure, composition, and function simultaneously, and at multiple scales. Because this kind of research is the most expensive to undertake, it is also becoming critical to build constituency support for scientific research. Finally, and perhaps most importantly, much controversy surrounding resource management questions is not due to the lack of science, or disagreements about the state of nature, but instead involve basic disputes about human values. Scientists and decision-makers need to make sure these issues are clearly separated in both the research and decision-making process.

Laws — The Good, the Bad, and the Ugly

Laws were identified both as facilitators and inhibitors of ecologically-based stewardship. Some saw the Endangered Species Act as a motivator for change to avoid the need for difficult and costly efforts to save individual species. Others encouraged creative use of the administrative flexibility in this and other acts of Congress to demonstrate more quickly the benefits of broad authority. Narrow mandates were seen generally to hinder problem resolution, because they raised the likelihood of litigation. Many reported that when stakeholders resorted to lawsuits, the chances for successful ecologically based stewardship diminished sharply as stakeholders withdrew in anticipation of the courts' command and control. Others pointed out that some laws simply were not cast to solve the current problems of cross-boundary and joint ownership activities on larger landscapes. Thus, jurisdictional issues were seen as needing clarification before new cooperative activities could be undertaken wholeheartedly.

I Need the Information... Yesterday!

Management decisions, and public education, are often hampered by the lack of current information from the scientific community. This is an important issue, especially given the dynamic nature of ecological stewardship issues. Participants reiterated the need to accelerate information dissemination through informal publications (that do not entail delays of months or years for publication), computer networks, and workshops that bring together the scientific and management communities. Scientists restated the need to ensure that their work was properly peer-reviewed, and implemented correctly. Continued thoughtful and up-front interaction between scientists and managers is the best approach to this challenge.

Broadening Horizons

Biological and biophysical scientists have a great deal to learn from interactions with social scientists, physicists, economists, and other fields not traditionally associated with "core" natural resource sciences. In addition, ecosystem scientists have to seek greater exposure to philosophers (for example, theologians and ethicists) to help describe appropriate directions for ecosystem science. And finally, ecosystem scientists need to gain insight from creative people working in fields such as literature, drama, movies and television, or photography. Such creative fields often reveal cultural preferences and concerns, and, in their own way, enable scientific ventures.

Can We Adapt?

Adaptive management, like ecosystem management, is one of those things that many people are already doing without realizing it. Participants felt, however, that "feedback loops" to provide managers with monitoring data and evaluations of their management actions were lacking in many agencies. Without this evaluation, managers have no way to assess the effectiveness of their actions and make necessary modifications. Many agencies have a history of collecting lots of data, but not of effectively using it for adaptive purposes. This was also recognized to be symptomatic of shifting agency goals. With the proper feedback loops, managers can be held accountable for their actions, and are, therefore, more likely to be responsive to local resource needs.

The Only Place it Matters Is on the Ground

No one at the Tucson workshop wanted to invest time in meetings, publications, or other activities that did not have a direct payoff for on-the-ground ecological stewardship. This was the underlying theme to every presentation and discussion. Clearly, the interactions between science and management authors led to a better understanding of what managers need and what scientists can do. This sort of interaction needs to become the norm, rather than the exception. Furthermore, the on-the-ground needs of the managers should strongly influence the research system.

CONCLUSION

Perhaps the most challenging aspect of ecosystem approaches to natural resources management is the need for interdisciplinary research and management. Context — environmental, social, and economic — is a backdrop for virtually all decisions in an ecosystem approach. Science is crucial, but reductionism is dangerous. To be effective, specialists often have to know something about issues they have little training or experience in. Economists working on management options for an area with endangered species may not need to be specialists in population viability analysis, but they should understand the basics of extinction processes in order to frame their analysis. Likewise, a restoration ecologist need not be expert with cost–benefit analysis to define a technical restoration plan, but a basic understanding of how economists evaluate costs and benefits will likely improve the feasibility of his or her proposal. And, resource managers often have little choice but to know something about many different issues — from building collaborative partnerships to knowing when to call for expertise in highly technical areas such as restoration ecology or legal advice. Ecosystem approaches simply will not work if researchers and managers isolate themselves from the context in which they work.

The topic summaries in Volume I are a good place to begin learning more about topics that you would like (or need) to know more about. The summaries in Volume I, however, are only previews that highlight findings from the full chapters in Volumes II and III. To make the most of this resource, we urge you to read those chapters for the full context of the topic. We are confident that the *Ecological Stewardship Reference* will help users to resolve specific problems and more fully understand the context for ecosystem approaches at the dawn of the 21st Century.

REFERENCES

Aplet, G.H., N.C. Johnson, J.T. Olson, and V.A. Sample. 1993. *Defining Sustainable Forestry*. Island Press, Washington, DC.

Cortner, H.J. and M.A. Moote. 1998. *The Politics of Ecosystem Management*. Island Press, Washington, DC.

Keystone Center. 1996. *The Keystone National Policy Dialogue on Ecosystem Management: Final Report*. The Keystone Center, Keystone, CO.

Leopold, A. 1949. *A Sand County Almanac*. Alfred Knopf Publishers, New York, NY.

Miller, K.R. 1996. *Balancing the Scales: Guidelines for Increasing Biodiversity's Chances Through Bioregional Management*. World Resources Institute, Washington, DC.

PCSD. 1996. *Sustainable America: a New Consensus for Prosperity, Opportunity, and a Healthy Environment*. The President's Council on Sustainable Development Final Report. President's Council on Sustainable Development, Washington, DC.

Sexton, W.T. 1998. Ecosystem Management: Expanding the Resource Management "Tool Kit". Special Issue on Ecosystem Management. *Landscape and Urban Planning*, 40: 103–112.

Sexton, W.T., N.C. Johnson, and R.C. Szaro. 1997. The Ecological Stewardship Project: A public–private partnership to develop a common reference for ecosystem management in the United States. In: *Proceedings of the XI World Forestry Congress: Forests, Biological Diversity and the Maintenance of Natural Heritage — Protective and Environmental Functions of Forests. Volume 2*. (Antalya, Turkey), pp. 75–82. United Nations Food and Agriculture Organization, Rome.

Szaro, R.C. and W.T. Sexton. 1996. Biodiversity conservation in the United States. In: A. Breymeyer et al. (eds.), *Biodiversity Conservation in Transboundary Protected Areas*, pp. 57–69. Proceedings of an International Workshop, Bieszczady-Tatry, Poland. National Academy Press, Washington, DC.

Thomas, J.W. 1979. *Wildlife Habitats in Managed Forests of the Blue Mountains of Oregon and Washington*. USDA Forest Service Handbook No. 533. USDA Forest Service, Portland, OR.

Yaffee, S.L., A.F. Phillips, I.C. Frentz, P.W. Hardy, S.M. Maleki, and B.E. Thorpe. 1996. *Ecosystem Management in the United States: An Assessment of Current Experience*. Island Press, Washington, DC.

Biological and Ecological Dimensions

♦ *Genetic and Species Diversity*

♦ *Population Viability Analysis*

♦ *Ecosystem and Landscape Diversity*

♦ *Ecosystem Processes and Functioning*

♦ *Disturbance and Temporal Dynamics*

♦ *Scale Phenomena*

♦ *Ecological Classification*

Genetic and Species Diversity

Why Genetic and Species Diversity Are Important To Ecological Stewardship

Human health and well-being fundamentally depend on a clean and steady supply of water and fertile soil to produce renewable resources for food and fiber. At the heart of the ecological processes that purify water, regulate its flow, build soils, recycle nutrients, pollinate crops, and decay organic waste is biodiversity, or the variety of species, genetic variability within species, and the ecological complexes in which species occur. Humans and all other species are utterly dependent on these "ecosystem services" and the species that individually and collectively perform these functions. Genetic and species diversity provide insurance against crop failures, maintain vital ecosystem processes and functions, and determine the resilience of ecosystems in the face of severe droughts, fires, and other disturbances. Sustainable resource management, therefore, requires understanding the factors that regulate species diversity, specifically those factors that either increase or decrease genetic and species diversity.

This summary was drafted by Robert Szaro with contributions from Michael Huston and Clive Steward. It is based on the following chapter in Ecological Stewardship: A Common Reference For Ecosystem Management, Vol. II:

Huston, M., G. McVicker, and J. Nielsen. "A Functional Approach to Ecosystem Management: Implications for Species Diversity."

KEYWORDS: biodiversity, habitat, ecosystem structure

The current store of genetic and species diversity is the foundation for all future forms of life on Earth. When genetic and species diversity are lost, so are opportunities to develop new and improved livestock and crop varieties, disease-resistant timber and shade trees, biological compounds for industrial processes, and pharmaceuticals for a range of human ailments from acne to AIDS, heart disease, and cancer. Unfortunately, the world is experiencing an accelerating loss of biodiversity — now the highest rate in at least 65 million years. The United States, itself home to more species than all but a few tropical countries such as Brazil, Colombia, and Indonesia, is experiencing significant losses. For example, The Nature Conservancy estimates that nearly a third of all species in the United States are at some risk of extinction. Understanding biodiversity and managing to protect genetic and species variation is in our best interest and at the core of ecological stewardship.

KEY FINDINGS

✓ **Biodiversity is an inherent property of all ecosystems**. Biodiversity is usually divided into three hierarchical categories — genes, species, and ecosystems (Fig. 1). Genetic diversity is the foundation of biodiversity at all levels, but most human interactions with biodiversity deal with individual species and species diversity. Species diversity is the total number of species of a particular type (e.g., canopy trees, understory plants, decomposing fungi, insectivorous birds) found within an area of a specific size.

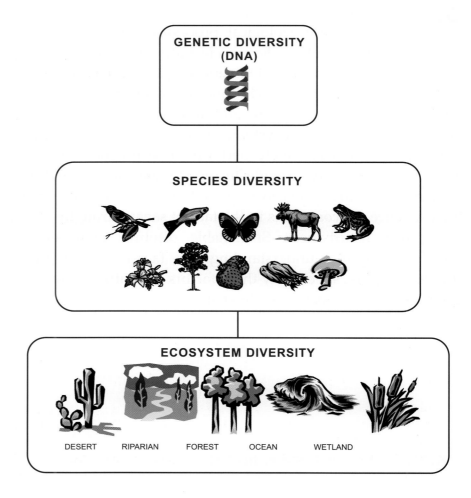

Fig. 1. Biological diversity is made up of all species of plants and animals, their genetic material, and the ecosystems of which they are part. (Images, copyright New Vision Technologies Inc.)

✓ **While biodiversity has many complex and interacting components, most management issues are only concerned with a small proportion of the total biodiversity**. In most situations, the total number of species is less important than ensuring an adequate level of the processes or functions that the individual species or groups of species perform. Identifying components of biodiversity on the basis of their function can provide a direct link to management objectives. However, given that the functional properties of many species are unknown, one should never assume that a particular species is redundant or unimportant.

✓ **Species diversity can change in response to both natural processes and human actions**. Fluctuations in species diversity are a natural phenomenon in all ecosystems, and occur in response to seasonal cycles, long-term climatic variations, and ecological processes such as succession. Extreme events, such as lightning strikes, landslides, and floods, can rapidly decrease species diversity, and alter environmental conditions so that future species diversity may be different than the diversity at the time of the event. Most management activities are either a cause of, or a response to, changes in environmental conditions that affect species diversity.

✓ **Changes in species diversity usually indicate that either physical or biological conditions have altered**. Changes in species diversity, as well as changes in the abundance or distribution of individual species, can indicate that conditions are changing. Different functional groups of species, however, may respond differently to the same environmental change. When a change in species diversity is observed, information on the particular species with the greatest response, whether diversity is increasing or decreasing, and changes in such properties as biomass or productivity, can help interpret what caused the change.

✓ **Species diversity is influenced by such physical conditions as climate, nutrient availability, physical structure or heterogeneity, and environmental disturbances that kill organisms**. The responses of species diversity to physical conditions are complex, but can be predicted if enough information about local conditions is available. A specific management action, such as increasing the harvest rate or adding fertilizer or water, can either increase or decrease species diversity, depending on the conditions under which the action is taken. Similarly, in response to any specific management action, the genetic diversity of some types of organisms is likely to increase, while that of the others will decrease. Knowledge of local conditions and how they may be changing is important for anticipating the impacts of management actions.

✓ **Biological interactions — such as competition, predation, mutualism, parasitism, and disease — also influence species diversity**. The importance of various biological interactions in regulating species diversity varies greatly. In stable and productive environments, the effect of biological interactions on species diversity tends to be stronger than in unstable or unproductive environments. In all ecosystems, biological and physical factors interact in many ways to determine the number and type of species present.

✓ **Impacts of management activities on species diversity should be evaluated in relation to the natural influences on species diversity**. Environmental heterogeneity, mortality-causing disturbances, and natural conditions that affect growth and productivity — drought for example — can intensify or moderate the impacts of management actions. The greatest threat to

**Box 1
Where Is Species
Diversity Highest?**

The diversity of many types of organisms is highest under conditions that are not particularly favorable for growth, such as dry, infertile soils, nutrient-poor waters, or otherwise unproductive conditions. Diversity is always low under extremely unfavorable conditions, but is often low under the most favorable conditions, where productivity is high and certain species of plants are very abundant. This pattern is strongest for plants in both aquatic and terrestrial environments, and is also found for the animals that depend directly on specific types of plant, such as insects. Other organisms, particularly predators, usually have their highest abundance and diversity under the most productive conditions, where their food resources are abundant. Consequently, neither nature nor resource managers can create high diversity of all types of species in a single type of environment.

**Box 2
How Can Species
Diversity Be Increased
Or Decreased?**

Resource managers can increase or decrease specific components of species diversity (e.g., functional groups such as understory plants, grassland herbs, small mammals, or birds) in three primary ways.

(1) Increasing the spatial structure and heterogeneity of the environment almost always *increases* overall diversity (both species diversity, and "landscape" diversity), but affects specific components of diversity that depend on the conditions created.

(2) Increasing the productivity of an environment, such as by adding resources like nutrients or water, will *decrease* plant diversity under most conditions. Resource addition can increase the diversity of either plants or animals when the supply of essential resources (nutrients or food) for those organisms is in very low supply, or when the frequency or intensity of mortality caused by predators or disturbances is high.

(3) Increasing the frequency or intensity of mortality-causing disturbances, such as mowing, harvest, or fires, will *increase* the diversity of plants under most conditions, except where productivity is very low, or where the frequency of disturbances is already high. Increasing the mortality rate for animals, by hunting or increased predation, will generally *decrease* animal diversity, except under the most productive conditions where a few species are very abundant.

biodiversity is habitat loss associated with the destruction, degradation, or fragmentation of specific habitats. These habitat changes do not necessarily eliminate all species, but change heterogeneity, disturbance dynamics, and resources so that some species are reduced or eliminated, and other species become more abundant. Unusual or chronic disturbances tend to favor invasive species over long-lived native species.

✓ **The invasion and spread of non-native species is a leading threat to genetic and species diversity on wildlands**. Examples are numerous and recognizable across the United States — from zebra mussels in the Great Lakes region, kudzu vine in the Southeast, leafy spurge on the northern Great Plains, and Gypsy moths in the Northeast. In some cases, introduced diseases have wiped out native species of tremendous ecological, social and economic importance. For example, American chestnuts — one of the most widespread and valuable species in eastern hardwood forests until chestnut blight was introduced from Asia — have disappeared from forests over their entire range. The spread of introduced (or "exotic") species is believed to be a close second to habitat loss as a factor in species extinctions and considerably more important than pollution and overharvesting. In addition, such non-native species can also reduce the level of many ecosystem functions valued by society, such as when leafy spurge invades native grasslands important for wildlife and livestock. Human transport facilitates the invasion of non-native species, both intentional and unintentional, across international and state borders. Human activities also help bring about the invasion of non-native species by making natural ecosystems more susceptible to invasion. Primarily, human activity makes ecosystems more

Fig. 2.
Purple loosestrife has become a widespread invasive species in wetland areas of the eastern United States.

susceptible to invasion by upsetting the natural community, particularly disturbing plant communities through plowing, mowing, harvesting, grazing, and trampling. Altering the natural pattern of resources, such as water and nutrients, can also facilitate the invasion of undesirable non-native species, or cause native species to become problems (Fig. 2).

PRACTICAL CONSIDERATIONS

All management actions could potentially affect species and genetic diversity. Genetic and species diversity will rarely be primary management objectives, except where threatened and endangered species occur, or in rare biological communities, such as tallgrass prairies, serpentine rock formations, or pristine streams. However, virtually all management actions, including the decision not to take action, can affect genetic, species, and even ecosystem diversity.

Sustainability is a key concept for conservation. Sustainable natural resource management, whether in forestry, fisheries, or agriculture, will help to conserve most components of biodiversity, including species and genetic diversity. Significant declines in species diversity are often a signal that natural resources are not being managed well. Declines in aquatic diversity, in particular, are leading indicators that both land and water resources are not being sustainably managed.

Proper land use planning can reduce conflicts. There is no inherent conflict between biodiversity conservation and sustainable resource uses, including agriculture and forestry. One reason is that the diversity of many organisms tends to be highest in marginally productive lands and waters that are not ideal for intensive human land use. Land use planning can assure that each part of the landscape is used for those purposes for which it is best suited. Steep or unproductive areas are more favorable for watershed protection and biodiversity conservation than they are for resource extraction. Fertile and relatively level areas are ideal for sustainable resource harvest, and tend to have lower species diversity, even under natural, undisturbed conditions. However, some components of species diversity, particularly large predators and other animals near the top of the food chain, have already been greatly reduced or even eliminated by converting most productive lands to resource production.

Both biological and physical data must be collected. Data should be collected on the distribution and abundance of species in conjunction with data on physical environmental conditions. The reason for this is that environmental conditions, such as disturbance regimes like fire, and resource availability such as water, strongly affect species diversity. High quality information on certain groups of organisms, such as plants, along with good data on physical conditions (soils, geology, slope, exposure, disturbances), can provide an adequate base of information for managing other species (e.g., insects) that are more difficult to sample.

Remote sensing can help quantify landscape features. Landscape features — such as hydrologic and geomorphological structures, geology, soil types, and the distribution of successional classes or forest management units — are most effectively quantified using aerial photography and satellite imagery. Geographic information systems (GIS) are well-suited to evaluating and measuring

Box 3

"There is a temptation to define one or a few species as 'essential' and the rest of the species that perform the same (or similar) function as 'redundant'. However, neither ecosystem processes nor the functional properties of species are understood sufficiently well to make this distinction with any confidence." (DiCastri, 1995).

these features, and for combining this information with field data from surveys and grid-based samples.

Models are an essential component of ecosystem management. Models are important because they can be used to evaluate alternative management plans, project future trends, and assess cumulative management impacts. Models for natural resource management planning range in size and complexity from flow charts to simple regressions to detailed computer simulations. Because multiple processes influence species diversity and ecosystem dynamics, models of tree growth and physiology, soil nutrient cycling and hydrology, plant succession and fire dynamics can all contribute to better ecosystem management.

SELECTED READING

Aber, J.D., and J.M. Melillo. 1991. *Terrestrial Ecosystems.* Saunders College Publishing, Philadelphia, PA.

Anderson, J.M. 1994. Functional Attributes of Biodiversity in Land Use Systems. In: D.J. Greenland and I. Szabolcs (eds.), *Soil Resilience and Sustainable Land Use.* Oxford University Press, Oxford.

Boyce, M.S., and A. Haney (eds.). 1997. *Ecosystem Management: Applications for Sustainable Forest and Wildlife Resources.* Yale University Press, New Haven, CT.

di Castri, F. 1995. Editorial. *Nature and Resources* 31 (3): 2.

Huston, M.A. 1994. *Biological Diversity: The Coexistence of Species on Changing Landscapes.* Cambridge University Press, Cambridge.

Mooney, H.A., and J.A. Drake (eds.). 1986. *Ecology of Biological Invasions of North America and Hawaii.* Springer-Verlag, New York, NY.

Oldfield, M. 1989. *The Value of Conserving Genetic Resources.* Sinauer, Sunderland, MA.

Oliver, C.D., and B.C. Larson. 1996. *Forest Stand Dynamics.* Second Edition. McGraw-Hill, New York, NY.

Scott, J.M., F. Davis, B. Csuti, R. Noss, B. Butterfield, G. Groves, H. Anderson, S. Caicco, F. D'Erchia, T.C. Edwards, J. Ulliman, and R.G. Wright. 1993. GAP Analysis: A Geographic Approach to Protection of Biological Diversity. *Wildlife Monographs* 123: 1–41.

Szaro, R.C., and D.W. Johnston (eds.). 1996. *Biodiversity in Managed Landscapes: Theory and Practice.* Oxford University Press, New York, NY.

Waring, R.D., and W.H. Schlesinger. 1985. *Forest Ecosystems: Concepts and Management.* Academic Press, New York, NY.

Population Viability Analysis

Why Population Viability Analysis Is Important To Ecological Stewardship

One of the dilemmas facing natural resource managers is how to meld the broad concepts of biodiversity and ecosystem management with the narrower focus of legislatively mandated management for individual threatened, endangered, or sensitive species. Ecosystem management, however, does not negate the need for research and management to preserve viable populations of individual species. Rather, ecosystem management reinforces the need to view individual species within the context of the ecosystems that sustain them. This means looking at species populations in terms of genetic and ecological levels of biological organization, understanding the role of species in ecosystem processes, and scaling management actions to ecological rather than political boundaries. Population viability analysis (PVA) is a valuable — and sometimes indispensable — tool in ecosystem management.

This summary was drafted by Nels Johnson with contributions from Barry Noon and Fred Samson. It is based on the following chapters in Ecological Stewardship: A Common Reference For Ecosystem Management, Vol. II:

Noon, B.R., R.H. Lamberson, L.L. Irwin, and M.S. Boyce. "Population Viability Analysis: A Primer on its Principal Technical Concepts".

Holthausen, R.S., M.G. Raphael, F.B. Samson, D. Ebert, R. Hiebert, and K. Menasco. "Population Viability Analysis in Ecosystem Management".

KEYWORDS: population viability, extinction, population dynamics, genetics

Box 1. What is Population Viability Analysis?

Population viability analyses are designed to assess the threats to a species' survival, and to help managers intervene before population declines become irreversible. PVA is a structured, systematic, and comprehensive examination of the interacting factors that place a population or species at risk. To estimate the likelihood that a population will persist for some chosen time into the future (often a hundred years), PVA requires the careful evaluation of such factors as reproduction rates, genetic characteristics, and geographic range requirements, generally by the use of mathematical analyses and computer simulations.

Population viability analysis is the study of all factors that may cause a species to go extinct (Box 1). It is important to ecological stewardship for several reasons. First, it is a relatively rigorous way to identify cause and effect relationships, and has accelerated our understanding of basic ecological relationships and underlying patterns in the distribution and abundance of living things. Second, PVA can help to monitor the health and integrity of ecosystems. For example, PVA for certain "keystone species" — organisms, such as salmon, that interact with a range of other species and vital ecosystem processes — may help to identify key factors or trends that affect the integrity of the entire ecosystem, providing an important contribution towards multi-species and ecosystem management. Third, by focusing on the factors most limiting to a species' survival, PVA can help to identify the most effective management actions to protect endangered species. Without well-executed PVAs, mistaken assumptions in species recovery plans can lead to the waste of precious time, money, and public support.

KEY FINDINGS

Although population viability analysis requires a sophisticated understanding of the biology of the species in question, a few basic concepts can clarify the role of PVA in ecological stewardship.

✓ **PVA grows out of a line of important developments in ecology**. The origins of PVA can be traced to work in the 1940s to define species-area relationships, and the biogeography of extinction and colonization of oceanic islands first developed in the 1960s. However, the most important antecedent of PVA is the early work on "minimum viable populations" (the size of a population below which extinction is likely) carried out by William Allee and others in the 1940s.

✓ **Many things can affect the persistence of a species, but not all of them are of equal importance**. Habitat loss, fragmentation, and the introduction of species, for example, are usually the most important risk factors in the decline of individual species. The most important factors, such as habitat loss, are called "deterministic" because they make populations of a species isolated and rare in the first place (Fig. 1). Secondary factors, such as severe weather events or predation, may contribute to the ultimate demise of a species but are rarely the primary cause.

✓ **Over the short term, deterministic factors such as habitat loss should get most of our attention**. Secondary factors, such as random fluctuations in population structure (demographic stochasticity), are usually symptoms or results of the deterministic factors. PVA can help to identify which deterministic factors have the most impact on a species.

Fig. 1. Isolated populations, such as remnant herds of bison at the turn of the last century, are particularly vulnerable to extinction. (Photo courtesy of US Fish and Wildlife Service.)

✓ **Simple statistics, such as population size and habitat area, are not enough to understand the ability of a species to persist**. It's crucial, in addition, to have a sense of the spatial and temporal relationships between habitat and demographic factors. PVA — whether quantitative or qualitative — should never be undertaken without a fundamental understanding of a species' ecology. Poorly executed PVAs can lead to misleading conclusions, and inappropriate management prescriptions. They may hasten rather than avert a species extinction, or alternatively lead to costly and unnecessary actions for a species that is not at high risk.

✓ **There is no single "best" approach for conducting a PVA**. The format for a given PVA is dictated by data availability, the degree to which the species' ecology and life history are understood, knowledge of risk factors, and management goals. Data availability is critical. It will determine whether a quantitative approach is possible, or whether it is better to choose a qualitative approach.

✓ **To be useful, PVAs should be directly linked to risk factors that are under the control of management**. Population viability analysis is useful only to the extent it helps to solve management problems. The results of a PVA should help managers decide on what actions to take — if any — to reduce threats to a species persistence. Actions might include control of invasive species, change in amount and shape of habitat, removal of toxic substances, etc.

The following points address when PVA should be used in the context of natural resource management, the basic requirements for a sound PVA, and limitations to PVA.

When:

✓ **PVA should be used when a species is believed to be on the road to extinction, but well before the point at which nothing can be done to save the species**. For extinct species such as the passenger pigeon, that point has come and gone (Fig. 2). Species listed under the Endangered Species Act

Fig. 2. Passenger pigeon. (Photo courtesy of US Fish and Wildlife Service.)

(ESA) in the United States are obvious candidates, and a new listing may trigger local demand for PVA by resource managers. PVA is often used to help determine whether a species should be considered for listing (and conceivably for de-listing as well).

✓ **A "coarse-filter" screen may be the best way to select species for PVA**. The best approach to selecting species for analysis may be to simply screen all species with in the planning area for viability concerns. A set of relatively simple criteria developed for the World Conservation Union (IUCN) Species Survival Commission can be used to determine which species are the best candidates for PVA. These criteria include population size, number, isolation, and habitat area trends. Other factors could include specific life history traits (e.g., reproductive rates), habitat specificity, and trends in abundance and distribution.

✓ **PVA may also be used as a pro-active tool**. For example, multi-species PVA is being used to help prioritize habitat restoration efforts and design monitoring systems in the Pacific Northwest. PVA could also be useful in making decisions about how to control invasive exotic species such as the zebra mussel or leafy spurge.

Basic requirements:

✓ **The most basic requirement is to know what data are needed**. This will depend on the species, area of analysis, and the management goal. Data on population size and dynamics, habitat area and dynamics, and risk factors are essential. Models used for PVA can help to explicitly identify what information is needed.

✓ **Determine whether necessary data exist**. For some species, the data may already have been collected by other researchers. For many species, some or most of the data will not exist. Information on population status, in particular, is key but available for only a limited number of species. Lack of data generally precludes use of quantitative approaches, but qualitative methods may still be useful. With either quantitative or qualitative methods, some data will likely have to be collected for the first time.

✓ **Basic information on population dynamics, genetics, and spatial and temporal dimensions of population change is essential.** There is no substitute for basic knowledge of population biology, genetics, ecosystem processes, and conservation biology. This knowledge is a prerequisite for conducting a PVA, even as software programs such as VORTEX, RAMAS, ALEX, etc. simplify the job.

✓ **Collaboration between researchers with both field and theoretical knowledge about a species is essential.** Between them, they should have knowledge of the species' habitat requirements, life history strategies, behavior (including breeding behavior and habitat selection), and dispersal.

✓ **PVA is not a casual undertaking.** Managers and researchers should appreciate the knowledge, money, time, and information needed to do a PVA correctly.

Limitations of PVA:

✓ **PVAs depend on knowledge that is often incomplete.** We do not have basic population and life history information on most species. For species where knowledge is limited, the risk of misleading results may outweigh the cost of rigorous PVA methodologies. Qualitative forms of assessment — expert-opinion workshops, for example — may be good alternatives. In some cases where information is extremely limited, it may be best to gather better data before any assessment is attempted.

✓ **PVA does not predict the fate of a species.** PVA is a probabilistic rather than a predictive tool. It focuses on the factors that are *most likely* to limit the persistence of a species over time.

PRACTICAL CONSIDERATIONS

It is not practical or possible to conduct a PVA for all — or even most —species in a given management situation. In general, it's advisable to focus on those species that are known or thought to be vulnerable to extinction or extirpation (Fig. 3). Species that may be good candidates for PVA include those currently listed under the Endangered Species Act as threatened or endangered, species ranked as G1, N1, or S1 in the TNC/Natural Heritage system, restricted range species (especially endemics) and species, including those not currently ranked as threatened or endangered, that are exhibiting chronic or precipitous population declines.

Six fundamental steps provide a basic framework — regardless of the method — for analyzing population viability in the context of ecosystem management:
1. Select species of concern;
2. Describe population status;
3. Describe risk factors;
4. Identify suitable habitat amount, distribution and trends;
5. Describe relationship of population dynamics to habitat dynamics;
6. Assess likelihood of species persistence.

Most examples of PVAs found in the published literature are atypical. Most published examples of PVA are for the relatively few species (<5%) for which we have good data. It is important to recognize that data for most species are sparse. This situation requires either a concerted effort to collect missing data (often a

Fig. 3. The Northern Spotted Owl has probably been the subject of more population viability studies than any other species in North America. (Photo courtesy of USDA Forest Service.)

costly and time-consuming proposition) or the use of more qualitative approaches.

Managers rarely have all the information needed to conduct a fully quantitative PVA. In these situations, managers must rely on simpler techniques that are less rigorous and possibly less reliable. Several alternative approaches to quantitative assessments include the use of expert opinion, habitat inventories, and basic information on current population status. With all methods, extensive documentation and use of a logical process are key to credibility.

Successful use of PVA depends on simple communication between the biologists performing the analysis and the managers using it. Biologists should understand that managers must look at an array of options, and that the role of analysis is to provide unbiased information on all reasonable options. Managers should understand the uncertainties involved in the analysis, and the limited ability of PVA to predict the future.

SELECTED READING

Boyce, M.S. 1992. Population viability analysis. *Annual Review of Ecological Systems* 23: 481–506.

Frankel, O.H, and M.E. Soulé. 1981. *Conservation and Evolution.* Cambridge University Press, Cambridge, UK.

Gilpin, M.E. and M.E. Soule. 1986. Minimum viable populations: processes of species extinction, pp. 19–34. In: M.E. Soule (ed.), *Conservation Biology: the Science of Scarcity and Diversity.* Sinauer Associates, Sunderland, MA.

Lande, R. 1993. Risks of population extinction from demographic and environmental stochasticity and random catastrophes. *American Naturalist* 142: 911–927.

Mace, G.M. and R. Lande. 1991. Assessing extinction threats: Toward a reevaluation of IUCN threatened species categories. *Conservation Biology* 5: 148–157.

Rabinowitz, D. 1981. Seven forms of rarity, pp. 205–217. In: H. Synge (ed.), *The Biological Aspects of Rare Plant Conservation.* Wiley, New York, NY.

Ruggiero, L.R., G.D. Hayward, and J.R. Squires. 1994. Viability analysis in biological evaluations: Concepts of population viability analysis, biological population, and ecological scale. *Conservation Biology* 8: 364–372.

Schonewald-Cox, C.M. 1983. Conclusions: Guideline to management: a beginning attempt, pp. 414–445. In: C.M. Schonewald-Cox, S.M. Chambers, B. MacBryde, and W.L. Thomas (eds.), *Genetics and Conservation: a Reference for Managing Wild Plant and Animal Populations.* Benjamin/Cummings, Menlo Park, CA.

Shaffer, M. 1981. Minimum population sizes for species conservation. *Bioscience* 31: 131–134.

Shaffer, M. 1987. Minimum viable population: Coping with uncertainty. In: M.E. Soulé (ed.), *Via Populations.* Cambridge University Press, New York, NY.

Shaffer, M.L., and F.B. Samson. Population size and extinction: A note on determining critical population sizes. *American Naturalist* 125: 145–152.

Soulé, M.E. 1987. *Viable Populations for Conservation.* Cambridge University Press, Cambridge, UK.

Ecosystem and Landscape Diversity

Why Landscape Diversity Is Important To Ecological Stewardship

Artists, architects, farmers, and tourists are all influenced by the landscapes in which they find themselves. So too, animals and plants are drawn to some parts of the landscape while avoiding others. Animals, plants, and the myriad natural processes that regulate ecosystems both respond to and help to create distinctive landscapes. Ecologists, of course, have a less romantic interpretation of landscapes than artists. Richard Forman, a leading landscape ecologist, defines a landscape as "a heterogeneous land area composed of a cluster of interacting components that is repeated in a similar format throughout" (Forman 1995). Understanding landscape diversity can help natural resource managers to respond effectively to diverse demands from wild species, loggers, canoeists, and even artists. It is at the landscape scale that managers have the flexibility to balance competing demands for natural resource extraction, recreation, and wildlife (Box 1).

KEYWORDS: landscape ecology, fragmentation, habitat patches, ecological classification, invasive species

This summary was drafted by Andrew Malk with contributions from Julie Concannon, James Gosz, and Henry Whittemore. It is based on the following chapters in Ecological Stewardship: A Common Reference for Ecosystem Management, Vol. II:

Gosz, J.R., J. Asher, B. Holder, R. Knight, R. Naiman, G. Raines, P. Stine, and T.B. Wigley. "An Ecosystem Approach for Understanding Landscape Diversity".

Concannon, J.A., C.L. Shafer, R.L. DeVelice, R.M. Sauvajot, S.L. Boudreau, T.E. DeMeo, and J. Dryden. "A Landscape Diversity Approach to Ecosystem Management".

In the Kisatchie National Forest in Louisiana, Forest Service staff turned to the Nature Conservancy and the Louisiana Wildlife and Fisheries Department to help them simultaneously plan timber harvests, conserve endangered species, and manage other resources. Together they developed an ecological classification. This classification uses geology, landform, climate, and other factors to identify potential natural vegetation — that is, the type of plant community that would be expected to grow there under natural conditions — at a landscape level. As in other parts of the South, conservation of the endangered red-cockaded woodpecker (*Picoides borealis*) (Fig. 1) has become a major management issue on the Kisatchie National Forest. The landscape ecological classification enabled forest staff to identify areas of longleaf pine that are prime habitat for the woodpecker. Based on this information, two red-cockaded woodpecker reserve areas were established. This also facilitated timber resource management since it helped clarify which areas were sensitive habitat areas and which were not.

Fig. 1. Red-cockaded woodpeckers. (Photo courtesy of US Fish and Wildlife Service.)

Managing for landscape diversity is a central feature of ecological stewardship for several reasons. First, a landscape approach makes it possible to identify the ecological processes, such as disturbance regimes, hydrologic and nutrient cycles, and biotic interactions, that are essential for maintaining biodiversity at a regional scale. Second, understanding landscape diversity can help identify areas that are least vulnerable to disruptive outside influences, such as pollution, introduced species, or creeping suburban sprawl. Third, a landscape approach to habitat conservation is important because isolated protected areas are rarely sufficient to provide enough habitat or resources needed to maintain a region's biodiversity over time. Assessing diversity at a landscape scale is essential to selecting reserves and planning resource management so that a full range of natural communities and their seral stages can be maintained.

KEY FINDINGS

✓ **Concepts of landscape ecology have evolved rapidly**. Common human perceptions are of diverse landscapes composed of croplands, farmsteads, roads, woodlands, and streams. The challenge for landscape ecology is to create a coherent body of knowledge based on patterns of species and habitat distribution and the frequency and patterns of ecological processes. As a separate field, landscape ecology considers the development and maintenance of spatial heterogeneity, interactions and exchanges across landscape boundaries, the influence of heterogeneity on biotic and abiotic processes, and the management of that heterogeneity (Box 2) (Fig. 2).

Fig. 2. Landscapes often encompass tremendous spatial heterogeneity, such as this landscape of riparian, forest, and meadow communities. (Photo courtesy of USDA Forest Service.)

✓ **Temporal and spatial scale are essential dimensions of landscape diversity**. A landscape may appear to be heterogeneous in appearance at one scale but homogeneous at another scale. Not all landscape processes occur simultaneously or at the same rate. Three mechanisms are largely responsible for landscape heterogeneity: (1) specific geomorphologic processes, such as erosion, occurring over long-time periods; (2) colonization patterns of organisms occurring over short and long time scales; and (3) local disturbances of individual ecosystems, such as fire or windstorms, occurring over a short time.

✓ **Landscape diversity is becoming more important in natural resource management decisions**. Many species, especially wide-ranging animals, use more than one habitat patch type and actively respond to the mosaic of habitats they encounter, avoiding some and using others. To understand landscape-level patterns and processes, it's important to know: where habitat boundaries occur and how they fluctuate; what determines where a boundary is located; how boundaries influence ecological processes within habitat patches and over the larger landscape; how boundaries affect the distribution and movement of materials, energy, and organisms across the landscape, and how these movements can change the location and nature of boundaries. Climate change and now human land uses are responsible for many of the most widespread and pervasive changes in the mosaic patterns of habitats and species that make up a landscape.

✓ **During the past century and more, many landscapes in the United States have shifted rapidly**. These landscapes, even those with moderate human occupation, have changed from a natural landscape matrix with human-altered habitat fragments embedded within it, such as extensive forests with farm clearings, to a managed landscape matrix with natural habitat

Box 2
Common Landscape Terms

Patch — A surface area differing in appearance from its surroundings.

Matrix — A landscape element surrounding a patch, also the most extensive and most connected landscape element present.

Fragment — A small remnant patch of a once larger habitat and typically isolated from other patches of the same habitat.

Landscape boundaries — Spatial discontinuities in features of the soil and/or vegetation; abrupt transitions between woodlands and grasslands, between riparian and desert vegetation, riparian vegetation and water, or between an alluvial fan and adjacent plateaus are clear examples.

fragments, such as agricultural fields and urban areas with scattered forest patches. Road building, logging, agriculture, and urban and industrial development tend to sharply fragment natural landscapes with abrupt boundaries and linear divisions (Fig. 3). Historical changes in land use and ecological processes have resulted in landscapes different in composition and structure from those that existed on the North American continent before European settlement. Understanding how landscapes have changed due to both natural and human processes can help managers anticipate how future natural and human changes will affect fire patterns, flood frequency, landslides, and forest and agricultural productivity.

✓ **Disturbances are fundamental for maintaining landscape and aquatic diversity**. Geological disturbances, including volcanism, continental and alpine glaciation, and erosion, fundamentally set the stage for landscape diversity. Climatic processes, including precipitation, seasonal changes, and warming and cooling cycles, help to determine which species and natural communities can survive in the landscape and where. Some ecosystems and natural communities require a certain level of disturbance (e.g., willow groves in riparian zones) to persist while others can be eradicated by unusually severe or frequent disturbances such as a catastrophic flood.

✓ **Riparian zones deserve special attention in maintaining landscape diversity**. Riparian communities, as interfaces between terrestrial and aquatic systems, encompass sharp environmental gradients, and link ecological processes and natural communities from land to water. Riparian communities are an unusually diverse mosaic of land forms, communities, and environments within the larger landscape. They are shaped by events

Fig. 3. This aerial photograph shows a mixed natural and human landscape in South Carolina with patches of natural areas interspersed with agriculture, housing, and roads. (Courtesy of Savannah River Ecology Laboratory.)

and processes across the landscape beyond their boundaries, and they in turn play a disproportionately important role in maintaining biodiversity at the landscape scale.

✓ **Individual species, populations, and natural communities can have a profound impact on landscape diversity**. In some landscapes, keystone species, such as beaver, can play a decisive role in ecosystem patterns and processes. Interactions between individual species and ecosystems are complex and difficult to identify. For example, population cycles of many keystone species can take place over decades. Research and monitoring is often over short periods of time and provide just a snap shot of interactions between species and ecosystems. Ecosystem changes are often subtle over the short term and shifts in biogeochemical cycles or soil formation, for example, may be detected only after several years or decades of monitoring. Nevertheless, landscape alterations by keystone species often cause patterns and processes to shift in ways that cannot be explained by longer term climate or geochemical processes. The landscape impacts of exotic species can often be seen more rapidly than the impacts of native species because they are typically aggressive dispersers and native species are unadapted to their presence. Invasive weedy species can affect landscape diversity on a grand scale. For example, leafy spurge, purple loosestrife, and a rapidly growing number of other invasive exotic plants can rapidly displace native species and fundamentally alter landscape diversity.

PRACTICAL CONSIDERATIONS

Managers need to address how internal and external forces affect ecosystem fragments in a landscape. Internal forces include tree mortality or changes in the population size of a keystone species. External forces include pollution, urbanization, introduction of exotic species, or climate change. Deciding which factors are more important often depends on the size of the ecosystem fragments. For large fragments, emphasis will be on the internal dynamics; for small fragments emphasis will be on the external influences. External factors, however, can be important regardless of the remnant size. Since most impacts on fragments originate from the surrounding landscape, managers need to shift away from traditional notions of management (single small patch management) toward integrated landscape management. Traditional natural resource management stops at the reserve boundary; however, fluxes of water, pollutants, and organisms do not. Placing important habitat fragments firmly within the context of the surrounding landscape and developing complement- ary management strategies is needed to ensure the long-term viability of remnant habitats.

Individual disturbance events should not be considered in isolation from other disturbance events. The impact of any disturbance event will be determined in part by the number and intensity of previous disturbance events. A modest disturbance, such as a small clearcut, may have an unusually large impact if it comes on top of a rapid succession of earlier disturbances.

Uniform landscapes are generally more vulnerable to the "amplification" of disturbances than heterogenous landscapes. A natural even-aged forest of lodge- pole pine in the northern Rockies and a large plantation of loblolly pine in the southeastern United States share similar vulnerabilities to catastrophic fire and disease because of their uniformity. Southern hardwood forests, with perhaps 30–40 tree species, may be less vulnerable to fire and disease.

Patterns at the landscape scale are more resistant to change than patterns at the scale of tree gaps, forest stands, and watersheds. However, when large-scale change does occur on a landscape, the recovery of the landscape patterns takes much longer than the recovery of more localized tree gap or forest stand patterns.

Landscape diversity consists of more than vegetation patterns. More than just vegetation patterns affect landscape diversity — geology (e.g., geomorphology and slope) and microclimate (e.g., humidity, wind direction, and speed) are also important factors. For example, at the stand level, clearcut logging creates forest patches with sharp edges. In southeastern Alaska, natural forest patches have gradual and transitional edges (e.g., muskegs and beachfront-forest edges). This difference in edge boundaries is ecologically significant. The microclimate conditions typical of interior forest environments — which affect plant reproduction and foraging quality for many interior dwelling mammals and birds — are found much closer to natural edges than clearcut edges in many cases. Thus, two adjacent areas with the same forest type and similar patch sizes, but different edge patterns, can have diverging ecological conditions that affect species composition, structure, and function.

Managers should look beyond relatively natural landscapes for conservation planning. Southern California, for example, is a biodiversity "hotspot" with varied ecosystems, rich species diversity, extensive habitat loss, and growing numbers of threatened and endangered species. In rapidly growing San Diego and Los Angeles counties, planners have approached the challenge of biodiversity conservation by thinking of the entire existing landscape as a *de facto* reserve system. They used gap analysis to survey several hundred thousand hectares and identify a set of core reserves, habitat corridors, and buffer zones that would most effectively meet their conservation objectives. Rather than limit their planning to undeveloped habitats, planners also examined how water reservoirs, golf courses, large estates, and even cemeteries could serve as habitat for fragmented populations, as stepping stones to larger habitat fragments, as sources of propagules for species reintroduction, and as safety nets in case of catastrophic disturbance, such as fires.

SELECTED READING

Chen, J., J.F. Franklin and T.A. Spies. 1992. Vegetation responses to edge environments in old-growth Douglas Fir forests. *Ecological Applications* 2: 387–396.

Johnson, N.C. 1995. *Biodiversity in the Balance: Approaches to Setting Geographic Conservation Priorities.* Biodiversity Support Program, Washington, DC.

Noss, A.F. and A.Y. Cooperrider. 1994. *Saving Nature's Legacy: Protecting and Restoring Biodiversity.* Island Press.

Shafer, C.L. 1994. Beyond park boundaries. In: E.A. Cook and H.N. Van Lier (eds.), *Landscape Planning and Ecological Networks.* Elsevier, Amsterdam.

Turner, M.G. and L.H. Gardner (eds.). 1991. *Quantitative Methods in Landscape Ecology.* Springer-Verlag, New York.

White, P.S., and S.T.A. Pickett 1985. Natural disturbance and patch dynamics: An Introduction. In: S.T.A. Pickett and J.N. Thompson (eds.), *The Ecology of Natural Disturbance and Patch Dynamics.* Academic Press, San Diego, CA.

Ecosystem Processes and Functioning

Why Ecosystem Processes and Functioning Are Important To Ecological Stewardship

Ecosystem processes and functioning offer a powerful approach to ecosystem management. Ecosystem management uses knowledge about ecosystem function and processes to obtain yields or services from ecosystems without irreversibly affecting their resilience and natural resistance to disturbances. Such management is termed sustainable: when it is conducted in an experimental mode with monitoring and re-adjustment, it is termed "adaptive management."

This summary was drafted by Robert Szaro with contributions from Ariel E. Lugo, Steven J. Paustian, Martha H. Brookes, and Don Imm. It is based on the following chapters in Ecological Stewardship: A Common Reference for Ecosystem Management, Vol. II:

Lugo, A.E., J.S. Baron, T.P. Frost, T.W. Cundy, and P. Dittberner. "Ecosystem Processes and Functioning."

Paustian, S.J., M. Hemstrom, J.G. Dennis, P. Dittberner, and M. H. Brookes. "Ecosystem Processes and Functions: Management Considerations."

KEYWORDS: ecosystem function, disturbance, succession, scale, riparian ecosystem, keystone species

For best results, functional and process-oriented approaches to ecosystem management require a hierarchical and functional classification of ecosystems. The best way to achieve such classification of ecosystems is by using environmental gradients as the scales along which organisms, populations, association, and life zones function. These environmental gradients form multidimensional ecological spaces that influence rates and direction of processes and functioning of ecosystems. The biota is the medium through which ecosystems harness, distribute, and accumulate energy and materials. Energy and materials flow through the complex web of feeding interactions within ecosystems and enter the ecosystem at critical interfaces with the atmosphere, soil, water, or other organisms.

Disturbances, whether natural or human-induced, influence ecosystem states over long periods. Ecosystem resilience in the face of disturbances, resistance to change, productive capacity, and ability to sustain productivity and complexity, all depend on maintaining ecosystem functions and processes. Organisms that perform critical functions in the ecosystem and are called "keystone species" or "natural ecosystem managers." Organisms that do not appear to have critical roles in the functioning of the ecosystem are said to have functions redundant with those of other organisms. When disturbed, ecosystems undergo succession or change that results in self-organization and adjustment to the new conditions.

KEY FINDINGS

✓ **Knowledge of ecosystem processes and functions underpins all ecosystem management principles**. This knowledge improves the ability of resource managers to balance the array of values that people demand from ecosystems, and is the only option for maintaining ecosystem sustainability, productivity, and integrity (Box 1).

✓ **Energy drives ecosystem processes, but energy use is not 100 percent efficient because of the laws of thermodynamics**. Because the supply of energy to an ecosystem is finite based on area and time, organisms have evolved to optimize the efficiency of energy and resource use within the constraints of their environment.

✓ **Ecosystems circulate, transform, and accumulate energy and matter through the medium of living activities**. The ecosystem is a functional entity with arbitrary geographical boundaries. The key processes of ecosystems include the following: biogeochemical cycles of water, sediments, rocks, and minerals; primary productivity; respiration; food–web interactions; and succession. These processes proceed at different scales of time and space, ranging from a single stand to the globe in the spatial dimension and from minutes to eons in the time dimension.

✓ **Natural disturbances continuously disrupt ecosystem processes and functioning and maintain most ecosystems in a constant state of biotic and environmental change**. Management causes additional changes that interact with natural disturbance regimes and together affect functions and processes of ecosystems at all scales of complexity (Fig. 1).

✓ **Sustainable land management requires an understanding of ecosystem functions and processes across a range of temporal and spatial scales**. Some of the approaches identified here for achieving the goal of ecosystem management through a functional perspective require understanding the

---

**Box 1
What is a process versus a function?**

A *process* is a sequence of events or states, one following from and dependent on another, which lead to some outcome.

Function in an ecosystem is the role that any given process, species, population, or physical attribute plays in the interrelations among ecosystem components or processes.

Fig. 1. Human disturbances such as this small dam have accelerated the flow of water and erosion. (Photo courtesy of USDA Forest Service.)

following: how ecosystem disturbances interact with ecosystem subsidies; how keystone species, also termed "natural ecosystem managers" function (Box 2), and the potential redundancy of function within different life-form components of ecosystems; how the initial condition of nutrients and organic matter in ecosystems affect the successional pathways of open and closed systems; how organisms, population, and ecosystems respond to different environmental gradients that result from human-induced or natural disturbances; what are the limits of ecosystem resilience and resistance to environmental change; and how to take advantage of the self-design properties of ecosystems.

✓ **Maintaining appropriate ecosystem diversity — species composition, structure, and pattern — sustains the system's resistance and resilience and reduces the likelihood of catastrophic change.** More cost-effective resource management will result from understanding ecological processes and functioning and, working with, rather than against, these processes and functions.

PRACTICAL CONSIDERATIONS

Resource managers can take advantage of the growing knowledge about ecosystem processes and functions to shape their management plans and actions.

Ecosystem functions vary with scale. Resource managers must determine the appropriate scales, indicators, and reference conditions when defining management objectives, inventorying, analyzing, manipulating, and monitoring ecosystems — and their processes and functions (Box 3). At the stand scale, resource managers address the production function directly through silvicultural techniques. Management actions at this level can be intensive. At the watershed scale, production is more complex to manage because of the heterogeneity of a watershed. A watershed may contain more than one stand, each with its own idiosyncrasies. Resource managers usually cannot manipulate production at regional and global scales, but they can influence the

**Box 2
Organisms as Ecosystem Managers**

The natural ecosystem manager is an organism that uses a small fraction of the total energy budget of the ecosystem but provides a key function that pushes the ecosystem in a certain direction. The functional outcome of the ecosystem manager's activities can be higher primary productivity, more efficient nutrient cycling, a service to other members of the ecosystem, or maintenance of the ecosystem in a particular state. The natural ecosystem manager may be inconspicuous and seemingly unimportant (for example, earthworms), dominant and destructive (for example, elephants), or its actions apparently aimed at a point far removed from the results (for example, predators that regulate the brouseline of forests by regulating herbivores).

Box 3
Four Scales of Ecological Complexity to Address Ecosystem Functioning

Stand — a system homogeneous in its biological, edaphic, climatic, or hydrologic conditions. Much of the silvicultural treatment of forests is based on stand treatments. Stands, which can be of any size, are open systems.

Watershed — a well-defined hydrological unit. A watershed, which can be small or immense, can contain one or many stands. It can include a whole landscape region or just part of one. Hydrological and geomorphological conditions define the watershed, rather than its biology, climate, geology, or soils. Watersheds provide well-defined boundaries for studying, managing, and conserving water, carbon, and nutrients.

Region or landscape — three components define a region or landscape: its visual or scenic component; its chronological component, that is, how its geomorphology developed; and the ecological component, that is, its ecological systems. Its boundaries are usually defined arbitrarily and represent a large area with a variety of open systems. A region or landscape may or may not have natural boundaries.

Global — the boundaries of the planet define the global scale. It is the only natural system almost closed to carbon and nutrients (meteorites account for a small input of mass from space) but not to energy.

---

global and regional production process through the cumulative effects of what is done at stand and watershed scales (Box 3).

Management is a process that adds or removes stress to ecosystems. Ecosystem managers, be they foresters, agronomists, or horticulturists, combine stress applications (disturbances) with stress removals (subsidies) to maintain ecosystem succession at a stage that matches their management objectives (Luken 1990). Applying or removing stressors is synonymous with "management practices". For example, pine forest managers in the southeastern United States commonly eradicate hardwoods, thus decreasing tree diversity and its associated ecosystem attributes. The primary productivity that would have been dissipated in maintaining (through respiration) the more complex hardwood systems is then concentrated as the woody biomass of the preferred pine forest. Because of the requisite costs, people continually seek to achieve management objectives at progressively lower costs.

Management is a form of ecosystem disturbance. The key objective in ecosystem management is maintaining ecosystem processes within the given range of natural variation or, in extreme cases, at least within acceptable thresholds that maintain ecosystem processes and functions critical to a functioning and sustainable ecosystem. Management differs from natural disturbances in that managing is a directed action with a purpose. Thus, the success of management depends on how well the disturbance is directed so that it achieves its purpose.

Analyses should be focused at the appropriate scale. When land managers are concerned about ecosystem process and function issues, ecosystem analyses should be focused at the watershed and landscape scales. The intensity of analysis, and which ecosystem functions appear to act in the system are functions of scale.

Using and determining reference points and conditions. The composition, structure, and landscape pattern of vegetation or other fundamental components of "natural" ecosystems provide a set of templates or reference points that have somewhat predictable influence on ecosystem process and function.

Managing within ecosystem constraints and using new management approaches. Using watershed analysis and adaptive management — and powerful tools for resource information gathering and analysis such as remote sensing and Geographic Information Systems — will increase the likelihood of meeting human needs and maintaining healthy ecosystems.

Understanding process and function is the key to effective management. Resource managers currently observe changes in many ecological processes and functions that result from their actions. By examining the causes of these changes, managers and others are creating new management approaches and policies designed to prevent or mitigate human interference with natural processes. By understanding and working with some of the ecological processes involved, managers can develop approaches that are effective in meeting many human needs and less disruptive to ecosystem integrity (Box 4).

Managing at the watershed scale. Federal agencies are widely implementing the Interagency Watershed Analysis procedure on federal lands across the Pacific Northwest Coastal Ecosystem. This process, which is designed to guide ecosystem analysis for watersheds of different sizes, consists of six steps to assess the dominant ecological processes, functions, conditions, and uses of a watershed. The analysis should result in a list of management recommendations, based on the goal of sustaining key ecosystem processes and functions. The analysis steps are simple, direct, and useful for a variety of scales and purposes (Box 5).

The importance of monitoring. The long-term effects on other ecosystem processes of managing a specific ecosystem process are often difficult to quantify and may be counterintuitive. To address these complex relations, ecosystem monitoring efforts must consider both the broad spatial scales and

---

Box 4
Western Rangeland Riparian Management

Background. Riparian ecosystems provide a good example of basic ecosystem functions, including nutrient capture, nutrient transfer, and material transfer. The key to understanding riparian ecosystem functions is knowledge of how fluvial processes (erosion and sedimentation) influence the transfer of energy and materials in a watershed and, in turn, shape streams and adjacent riparian vegetation communities. Fluvial erosion is the dominant process for transfer of sediments from headwaters to valley stream segments. Vegetation in headwater zones acts to mediate rain splash and surface erosion. It also facilitates infiltration of precipitation that feeds shallow groundwater aquifers and is ultimately released into springs and surface-water bodies. Riparian areas are mainly influenced by sediment deposition during over-bank floods. Floodplain deposition zones act as sinks for both sediments and nutrients. These deposits also slow the release of groundwater from stream terraces that sustain stream flow during droughts. The availability of nutrient-rich soils and water make productive sites for vegetation adapted to periodic flood disturbances. The roots of riparian vegetation also aid in stabilizing stream banks and create cover from overhanging banks. Riparian vegetation has another important function in trapping new sediments and reducing streambank erosion during over bank flood events. Extreme flood events recycle soils through floodplain erosion and transfer organic and inorganic sediment to downstream deposition sites.

Lessons learned from past riparian management. Rangeland riparian areas represent a small portion of the landscape in the West, usually less than one percent of the land area (Elmore and Beschta 1987). In eastern Oregon, the interaction of trapping beaver and livestock grazing has altered basic ecosystem functions by changing the rates of sediment routing and storage in streams, reducing nutrient capture and transfer in riparian and aquatic ecosystems, and changing the hydrology of many stream systems. Concentrated, year-round foraging by domestic livestock has caused expanded gully networks, caused ephemeral flow regimes in streams that once had perennial flow, simplified stream banks and channels, and changed willow and wet meadow riparian communities to more xeric plant types. Grazing has also induced changes to riparian areas, which intensified the effects of natural disturbances and, in many instances, inhibited recovery from extreme flood events.

Successful management actions that have restored natural processes and improved rangeland riparian functions. Management actions included changes to grazing patterns in degraded watersheds and long-term exclusion of grazing in sensitive riparian areas. Grazing restrictions for periods of 5 to 20 years resulted in dramatic improvements in riparian vegetation health and diversity.

Box 5
Six Steps for Watershed Analysis

(1) Characterize the watershed. Identify the dominant physical, biological, and human processes of features that affect ecosystem functions or conditions.

(2) Identify issues and key questions. Surface the key elements (indicators) of the ecosystem that are most relevant to management questions and objectives, human values, or resource conditions within the watershed.

(3) Describe current conditions. Develop information relevant to the issues and key questions identified in step 2.

(4) Describe reference conditions. Explain how ecological conditions have changed over time as a result of human influences and natural disturbances.

(5) Synthesize and interpret information. Compare existing and reference conditions of specific ecosystem elements and explain significant differences, similarities, or trends and their causes.

(6) Formulate recommendations. Bring the results to conclusion, focusing on management recommendations that are responsive to watershed processes identified in the analysis.

---

the long-term temporal scales in which these processes operate. Because individual site management programs may influence or be influenced by these broader processes, managers must collaborate to coordinate monitoring at the landscape scale. The purpose of coordinated monitoring is to provide early warning about landscape trends that may influence the freedom or appropriateness of individual actions. Landscape-scale processes might negate a positive outcome at a given site; or one project's effects might compound the unwanted outcomes of another manager's programs.

SELECTED READING

Brown, S., and A.E. Lugo. 1994. Rehabilitation of tropical lands: A key to sustaining development. *Restoration Ecology* 2 (2): 97–111.

Evans, F.C. 1956. Ecosystem as the basic unit in ecology. *Science* 123: 1127–1128.

Hamblin, W.K. 1985. *The Earth's Dynamic Systems*. Burgess Publishing, Minneapolis, MN.

Lugo, A.E. 1978. Stress and ecosystems. In: J.H. Thorp and J.W. Gibbons (eds.), *Energy and Environmental Stress in Aquatic Systems*. U.S. Department of Energy Symposium Series CONF 77114. National Technical Information Services, Springfield, VA.

Lugo, A.E. and C. Lowe (eds.). 1995. *Tropical Forests: Management and Ecology*. Springer Verlag, New York.

Luken, J.O. 1990. *Directing Ecological Succession*. Chapman and Hall, New York.

Majer, J.D. (ed.). 1989. *Animals in Primary Succession: The Role of Fauna in Reclaimed Lands*. Cambridge University Press, Cambridge, UK.

Matthews, J.D. 1989. *Silvicultural Systems*. Clarendon Press, London.

McIntosh, R.P. 1985. *The Background of Ecology; Concept and Theory*. Cambridge University Press, Cambridge, UK.

Mills, L.S., M.E. Soul, and D.F. Doak. 1993. The Keystone Species Concept in Ecology and Conservation Biology. *BioScience* 43: 219–224.

Morrison, P. and P. Morrison. 1982. *Powers of Ten: About the Relative Size of Things in the Universe*. Scientific American Books, New York.

Naiman, R.J. (ed.). 1992. *Watershed Management: Balancing Sustainability and Environmental Change*. Springer Verlag, New York.

Odum, H.T. 1983. *Systems Ecology: An Introduction*. John Wiley & Sons, New York.

Odum, H.T. 1996. Scales of ecological engineering. *Ecological Engineering* 6: 7–19.

O'Neill, R.V., D.L. DeAngelis, J.B. Waide, and T.F.H. Allen. 1986. *A Hierarchical Concept of Ecosystems*. Princeton University Press, Princeton, NJ.

Schlesinger, W.H. 1991. *Biogeochemistry: An Analysis of Global Change*. Harcourt Brace Jovanovich Publishers, Academic Press, San Diego, CA.

Disturbance and Temporal Dynamics

Why Disturbance and Temporal Dynamics Are Important To Ecological Stewardship

Natural disturbances — fires, floods, wind and ice storms, and insect and disease outbreaks — shape ecosystems. Species respond to disturbances over ecological and evolutionary time by developing traits and structures that reflect the influence of disturbance. In fire-prone climates, for example, many trees have thick bark, serotinous cones, or basal sprouting that help them survive and reproduce following fire. Disturbances also mold community structure. Ice storms snap canopy trees, or wave action during storms may dramatically crop kelp beds.

This summary was drafted by Robert Szaro with contributions from Peter S. White, R. Todd Engstrom, Jonathan Harrod, Greg Aplet, and Beverly Collins. It is based on the following chapters in Ecological Stewardship: A Common Reference For Ecosystem Management, Vol. II:

White, P.S., J. Harrod, W.H. Romme, and J. Betancourt. "Ecosystem Disturbance and Temporal Dynamics."

Engstrom, T., S. Gilbert, M. Hunter, D. Meriwether, G. Nowacki, and P. Spencer. "Practical Applications of Disturbance Ecology to Natural Resource Management."

KEYWORDS: landscape patterns, climate change, ecosystem processes, fires

Fig. 1. Crown fires are extremely difficult to control and cause significant tree mortality. (Photo courtesy of USDA Forest Service.)

Humans can alter the frequency and severity of disturbance. The introduction of Eurasian cheatgrass to western rangelands, for example, has resulted in increased fire frequency and loss of native cover. Suppression of natural disturbances can also lead to changes in ecosystem composition and structure that threaten ecosystem integrity. In giant sequoia groves, for example, suppression of small, low-intensity fires leads to increases in fuel levels and larger, less controllable fires (Fig. 1).

Many management actions — intentional disturbances of ecosystems — affect the natural disturbance regime. These interactions cause many of today's most vexing natural resource management issues. Obvious examples include: channelizing rivers and eliminating floodplain vegetation that causes highly destructive flooding, fire suppression in pyrogenic vegetation types that results in wildfires, and logging on steep terrain that causes mudslides and river siltation. Effective ecosystem management requires a working knowledge of the role and characteristics of disturbance in ecosystems. Natural disturbance regimes can provide conservative guidelines for sustainable management on such issues as silvicultural practices, dam releases, and fire management policies. Understanding how natural and human-caused disturbances interact is critical to a manager's ability to meet land stewardship objectives, including the maintenance of biodiversity, aesthetics, air and water quality, and commodity production.

KEY FINDINGS

✓ **Disturbances are discrete events that change ecosystem structure and resource availability.** Disturbances differ in spatial characteristics such as size and pattern, temporal characteristics such as frequency and seasonality, intensity (percent of biomass killed or removed), and which species they affect. All these characteristics, considered together, define a disturbance regime (Box 1).

✓ **Although some disturbances kill only a small percentage of living biomass, others may remove most or all of the organisms within a patch.**

The amount of living and dead material left in a patch after disturbance determines community and ecosystem response. This "biological legacy" may include standing live trees, seed banks, standing snags, logs and other coarse woody debris, humus layer, and soil biota. These residual organisms and structures may help maintain ecosystem function, moderate fluctuations in temperature and humidity, and restrict nutrient and sediment loss during the early stages of post-disturbance recovery.

✓ **Disturbances are dynamic processes responsible for change in pattern on the landscape over time**. Disturbance processes create landscape patterns; these patterns, in turn, control subsequent processes. For example, the practice of "checkerboard" cutting in the forests of the Pacific Northwest produces a pattern of small clearcuts dispersed across the landscape. Subsequent windthrow damage is concentrated along the edges of clearcut patches. Compared with a landscape with a few large patches, a landscape with many small patches has a greater length of edge per area of clearcut. Thus, dispersed cutting creates a pattern (many small patches with high total edge length) that promotes a disturbance process (windthrow along forest edges). The relationship between landscape patterns and disturbance processes has important implications for ecological stewardship.

✓ **The nature of both the pattern and the processes a disturbance creates will differ with scale**. Details are lost but broader patterns emerge over areas of increasing size. At the scale of a few hundred square meters, the spatial pattern of tree crowns in a southern Appalachian forest depends on the birth and death of individual trees. Over many hectares, single trees are no longer visible; the pattern of stands reflects minor landforms and stand-level disturbance history. At the scale of tens of square kilometers, individual stands merge into larger land-cover units, and vegetation patterns follow broader physiographic gradients and boundaries between land ownerships. There is no one scale that is best for all management applications; the scale used will depend on the questions asked and the resolution of the available data.

✓ **Ecosystems are affected by multiple disturbances, and combinations of disturbance produce unique effects**. In tallgrass prairie, disturbance interactions promote diversity at both patch and landscape levels. Fire and grazing create four patch types: undisturbed, grazed, burned, and both burned and grazed. These patch types differ in vegetation structure and species composition. Burned, ungrazed patches are least diverse while burned, grazed patches are most diverse. Fire and grazing act together to maintain variety of patch types in the landscape, and high levels of diversity within patches.

✓ **The history of disturbance on a site can influence the outcome of subsequent disturbances**. For example, fires alter chaparral soils and increase likelihood of landslides on steep slopes. Feedbacks and interactions have been documented in many systems. The widespread occurrence of feedbacks and interactions suggests that disturbances cannot be studied or managed as independent events.

✓ **Natural disturbances share some characteristics with anthropogenic disturbances, but may differ in important ways**. For example, clearcutting and windthrow both remove the forest canopy, but clearcutting removes much of the woody biomass that would remain after windthrow. Managers may capitalize on such differences; for example, they may clearcut to

---

Box 1
Describing disturbance regimes

Kind:
- What kinds of disturbances are present in the system?

Spatial characteristics:
- What is the range of disturbance sizes and shapes?
- How does the occurrence of disturbances vary with environment and vegetation?

Temporal characteristics:
- How often do disturbances occur?
- What is the average return time for disturbances?
- During what seasons do disturbances occur?
- How do weather patterns and year-to-year climate fluctuations affect disturbance?

Specificity:
- What species and size classes are affected?

Magnitude:
- How much living biomass is killed?
- What organisms and structures are left behind by disturbance?

Synergisms:
- What are the effects of a disturbance on subsequent disturbances?

Fig. 2. Logging prescriptions can be designed to mimic the dynamics of certain natural disturbances, such as retaining scattered live trees as habitat patches. (Photo courtesy of USDA Forest Service.)

promote vegetation that regenerates best under open conditions. Alternatively, managers may seek to mitigate impacts by emulating characteristics of natural disturbance. For example, retention of scattered live trees and coarse woody debris following logging may provide wildlife habitat and sites for vegetation establishment (Fig. 2).

✓ **The effects of fire suppression on fire size and intensity have become management issues in many landscapes.** In chaparral and lodgepole pine, where crown fires dominated the historical disturbance regime, recent fire suppression has not fundamentally altered landscape structure or dynamics. Fire suppression in sequoia-mixed conifer forests in California and ponderosa pine woodlands in the American West, which were maintained historically by frequent low-intensity surface fires, has led to increases in stand density and fuel loads, creating the conditions for high intensity crown fires — Yellowstone was a lodgepole pine ecosystem!

Fig. 3. Roads are the major source of severe erosion and landslides in many steep areas. (Photo courtesy of USDA Forest Service.)

✓ **Relationships between landscape heterogeneity and disturbance are complex.** Some disturbances originate at forest edges; for example, landslides and human-caused ignitions are often associated with roads (Fig. 3). Large, homogeneous tracts of susceptible habitat may facilitate the spread of fires and insect outbreaks.

✓ **Compositional change is a feature of the normal dynamics of many systems.** Disturbances such as fires and insect outbreaks can spread over large areas, and are often linked to regional climatic patterns. For these reasons, even very large management areas may show considerable fluctuations. Intervention may be necessary when management units are not large enough to retain all species and successional stages, or when large fluctuations in landscape pattern may interfere with management objectives such as constant, sustainable levels of harvest.

✓ **Climate change, the pervasive influence of human cultures, and long intervals between some major disturbances make it difficult to fully document disturbance regimes.** Because disturbance regimes are ultimately coupled to climate, differences between past and present climates may limit the application of historic disturbance data to present situations. Human-induced changes in climate and atmospheric chemistry are likely to further complicate the situation in the future. Historical and prehistorical land-use have affected disturbance regimes in both direct (burning and clearing) and indirect (e.g., changes in landscape structure) ways. Disturbance regimes are not necessarily stable, and any demonstration of stability is contingent on temporal scale. For these reasons, we should be careful in our interpretation of and reliance on "natural," "original," or "historical" (e.g., pre-settlement) disturbance regimes.

✓ **The spatial and temporal characteristics of a disturbance regime have important implications for the existence and nature of equilibrium.** Steady-state equilibrium is most likely to occur where disturbances are small and infrequent relative to recovery time and total landscape size. While landscapes with large or frequent disturbances may not be in quantitative equilibrium, they may nonetheless exhibit qualitative or stable-trajectory equilibrium. Habitat loss and fragmentation alter the scales of landscapes, making it more difficult to preserve species and maintain historical disturbance regimes.

✓ **Before applying disturbance ecology, managers must investigate the relationship of various factors.** These factors include the relation between climate and disturbance regime, the effects of Native Americans, the influence of habitat fragmentation on the spatial and temporal characteristics of disturbance, the relationship between exotic species invasions and disturbance, and the relation between ecological variation and resilience.

PRACTICAL CONSIDERATIONS

There are several limitations on understanding and incorporating disturbance regimes: (1) Sources of information on historical disturbance are limited to cases that leave a permanent record (e.g., fire scars, photographs). Many kinds of disturbance in many ecosystem types leave no permanent record. (2) Characterizations of historical variation due to disturbance are sensitive to the reference period and spatial scale considered. (3) Many systems have been fundamentally altered due to species extinctions, species invasions, and changes in hydrology. Climate change may produce conditions that alter future disturbance regimes.

Restoring natural processes may first require restoring structure and pattern. For example, in areas where fire suppression has led to increases in stand density and large fuel buildups, thinning treatments and fuel-reduction pre-burns may be necessary before low-intensity surface fires can be reintroduced. In some systems, several cycles of disturbance may be required to restore historical structure.

Monitoring disturbance events and their impact is essential. This is true whether disturbances were initiated as part of an ecosystem management plan, or were a natural or accidental event (Fig. 4). Without rigorously designed and executed monitoring there is no way to evaluate the efficacy of management and take corrective action if necessary (adaptive management). Over the long term, monitoring will lead to a deeper understanding of ecosystem dynamics and more effective ecosystem management.

Setting goals in changing environments: We have to recognize that most biological communities are always recovering from the last disturbance. In the context of ecosystem management, goals represent the desired end state, and reflect policies and values of the organizations and individuals who define them. However, the concept of setting goals may appear futile in an ever-changing system. How can we be assured that our goals for natural resource conditions and uses are attainable and sustainable? How can we reduce the risk of failure in achieving goals, given the unpredictable nature of disturbance events and other dynamics? These questions can be addressed by setting goals with a historical perspective of inherent variability within the systems, and within the context of current dynamics that continue to shape natural resources. The range of variability and the nature of the current forces that affect ecosystems help us to define ecological limits that determine the range of conditions around which human desires and expectations should be shaped.

Fig. 4. Monitoring patterns of natural disturbances can provide managers with ideas for modifying management practices. (Photo courtesy of USDA Forest Service.)

SELECTED READING

Agee, J.K. 1993. *Fire Ecology of Pacific Northwest Forests.* Island Press, Washington, DC.

Attiwill, P.M. 1994. The disturbance of forest ecosystems ecological basis for conservative management. *Forest Ecology and Management* 63: 247–300.

Christensen, N.L., A.M. Bartuska, J.H. Brown, S. Carpenter, C. D'Antonio, R. Francis, J.F. Franklin, J.A. MacMahon, R.F. Noss, D.J. Parsons, C.H. Peterson, M.G. Turner, and R.G. Woodmansee. 1996. The Report of the Ecological Society of America Committee on the Scientific Basis for Ecosystem Management. *Ecological Applications* 6: 665–691.

Holling, C.S. 1995. What barriers? What bridges? In: L.H. Gunderson, C.S. Holling, and S.S. Light (eds.), *Barriers and Bridges to the Renewal of Ecosystems and Institutions.* Columbia University Press, New York, NY.

Hunter, Jr., M. 1990. *Wildlife, Forests, and Forestry: Principles of Managing Forests for Biological Diversity.* Prentice Hall, Englewood Cliffs, NJ.

Kaufmann, M.R., R.T. Graham, D.A. Boyce, Jr., W.H. Moir, L. Perry, R.T. Reynolds, R.L. Bassett, P. Melhop, C.B. Edminster, W.M. Block, and P.S. Corn. 1994. An Ecological Basis for Ecosystem Management. USDA Forest Service, General Technical Report RM-246.

Kohm, K.A. and J.F. Franklin (eds.). 1997. *Creating a Forestry for the 21st Century.* Island Press, Washington, DC.

Sousa, W.P. 1984. The role of disturbance in natural communities. *Annual Review of Ecology Systematics* 15: 353–391.

Swanson, F.J., and J.F. Franklin. 1992. New forestry principles from ecosystem analysis of Pacific Northwest forests. *Ecological Applications* 2: 262–274.

Swetnam, T.W. and A.M. Lynch. 1993. Multicentury, regional-scale patterns of Western Spruce Budworm outbreaks. *Ecological Monographs* 63: 299–424.

Turner, M.G. 1989. Landscape ecology: the effect of pattern on process. *Annual Review of Ecology and Systematics* 20: 171–197.

White, P.S. 1979. Pattern, process, and natural disturbance in vegetation. *Botanical Reviews* 45: 229–299.

Scale Phenomena

Why Consideration of Scale Is Important To Ecological Stewardship

Research and management of natural resources carried out at a single scale has a history of unpleasant surprises. For example, research in the 1950s and 1960s showed that the richness of local species and the abundance of game species generally increased in proximity to habitat edges. This prompted wildlife biologists and other natural resource managers to recommend increasing edge areas through small dispersed clearcuts. Landscape scale analyses in the 1980s, however, led researchers and natural resource managers to a different conclusion — creating lots of edge may increase local species diversity, but the resulting habitat fragmentation may lead to a net loss in biodiversity at the landscape level. The realization that analyses at different scales can lead to significantly different conclusions has spurred the rise of ecosystem management approaches (Box 1).

This summary was drafted by Andrew Malk with contributions from David Caraher, John Haufler, Jan Sendzimer, and Bob Szaro. It is based on the following chapters in Ecological Stewardship: A Common Reference For Ecosystem Management, Vol. II:

Caraher, D.L., A.C. Zack, and A.R. Stage. "Scales and Ecosystem Analysis."

Haufler, J.B., T. Crow, and D. Wilcove. "Scale Considerations for Ecosystem Management."

KEYWORDS: landscape, spatial resolution, temporal scale, planning

> **Box 1**
> **Ecosystem Management is Impossible Without Scale Considerations**
>
> Ecosystem management approaches typically involve a range of ecological, social, and economic objectives. No one scale for analysis, planning, and management is appropriate for all of these objectives. The geographic range needed for a population of grizzly bears to survive will be much greater than for a white-tailed deer. Natural resource extraction may satisfy a local community's economic objectives while a region's economic growth may depend more on natural amenity values that attract investments and new residents. An ecosystem management project can take place at a range of geographic scales — from a few thousand acres to several million acres. Regardless of the project area, the scale will determine the issues and opportunities to address. At each scale, one needs different information and management tools to meet management objectives. Institutional jurisdictions and stakeholders may vary as well. Not only geographic scales, but temporal scales help define ecological, social, and economic systems. For example, some forest types have a much more rapid rate of successional change than others. If one management objective is to maintain a full range of successional classes, this rate becomes critical in determining appropriate harvest rotation periods. Identifying appropriate geographic and temporal scales for analysis, planning, and management is indispensable to ecosystem management.

Why is it important to understand the relationship between scale and ecosystem management objectives? First, ecosystems are not closed, self-supporting systems, but rather are parts of larger interacting systems. This is also true of social and economic systems. For example, when surveying the economic and social objectives of an area, local community interests may be different from state or national level interests. Second, various ecosystem elements (e.g., species, soils) and processes (e.g., fire regimes, nutrient cycling) exist at different scales (Fig. 1). Third, regardless of the particular issue or question, there is always need for a larger scale perspective to deal with cumulative impacts and to establish context and framework for actions. Lack of a larger perspective is the most common scale-related problem in forest and natural resource management.

Successful ecosystem management relies on identification of appropriate scales and their linkages. Larger scale ecosystem and landscape perspectives are relatively new, and tools for assessing larger scales and incorporating them into scientific decision making are underdeveloped. Language describing phenomena and processes (biological and social) in terms of their spatial and temporal scales is still emerging, but shows promise in addressing complex problems which involve human and natural systems.

KEY FINDINGS

No single scale can meet all objectives of ecosystem management. Therefore, consideration of scale is critical to ecosystem management efforts. Scale considerations should include the following.

✓ **Determining the boundaries of the planning landscape**. Once management objectives are defined, the next step is to determine the extent and boundaries of the planning landscape. The spatial coverage of the planning landscape must be large enough to address population viability, biodiversity, or other factors addressed by planning objectives. However, the planning landscape should be small enough to include only as much

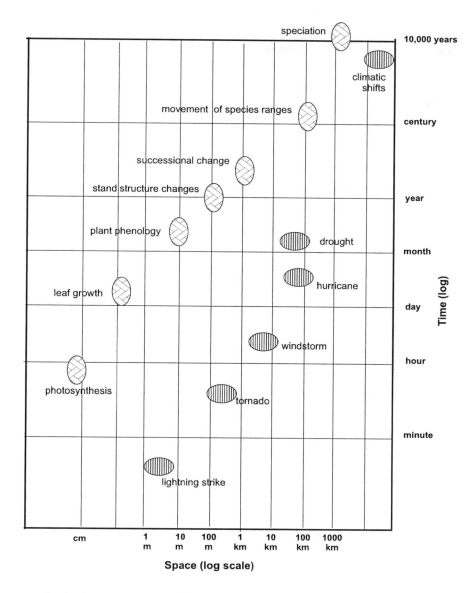

Fig. 1. Ecosystem processes exist at a range of spatial and temporal scales (adapted from Holling 1995).

ecological variation as needed to address those objectives and to ensure it is practical for developing feasible collaborative partnerships and can make use of existing information and databases.

✓ **Determining spatial and temporal resolution to address management objectives**. The resolution of mapping units and data can significantly influence the conclusions of an ecosystem analysis. Typically, management objectives are related to various ecosystem processes, each with their own unique scales in space and time. It is critical that the data used can detect the relevant processes and phenomena. For example, if a management objective is to base silvicultural practices on natural disturbance regimes, it may be necessary to use fine resolution data to detect tree-fall gaps while coarse scale data is adequate for detecting fire frequency and distribution patterns. Spatial resolution should be set by the minimum size of the object or phenomena of interest. The appropriate temporal resolution should be determined by the shortest increment of time needed to detect the periodicity of the process or phenomena of interest. For example, the impact of recreational rafters on a river system can be gauged more accurately with monthly or seasonal data than with annual data.

✓ **Determining the relevant planning horizon.** Some management objectives can be safely achieved with a relatively short planning timespan or horizon while others cannot. Given that ecosystems typically have successional trajectories that last for decades or even centuries, long-term planning horizons are frequently needed. On the other hand, social and economic conditions rapidly change and relatively short-term planning horizons may be more appropriate. In some cases, it may be appropriate to set long-term planning horizons for certain objectives, while setting shorter horizons, or periodically revising parts of the long-term plan, for others.

✓ **Assessing an appropriate timespan for an historical perspective.** Historical reference points can help set ecological objectives. An appropriate timespan for an historical perspective on ecological conditions will be determined by the period of interest. If the objective is restoring "natural" ecological conditions, that period might have to include reconstruction of disturbance events over several centuries prior to the arrival of humans in North America ten or twelve thousand years ago. If the goal is pre-European conditions, the time span would need to be set differently, perhaps between 1000 and 1500 A.D.

PRACTICAL CONSIDERATIONS

Selecting the appropriate scales of analysis often challenges resource professionals. The analytical scale should be selected to match its purposes (Table 1). For example, a district ranger designing a 40 acre timber harvest might consider a planning area limited to a watershed of several thousand

Table 1. Analytical considerations for three broad spatial and temporal scales.

Analytical Considerations	Relative Scale		
	Large	Mid	Small
Purpose or interest	Develop policies, laws, standards, and practices	Develop management strategies and programs	Design projects
Spatial scale	> 2,000,000 acres	2,000–2,000,000 acres	< 2,000 acres
Temporal Scale	> 100 years	10–100 years	1–10 years
Focus	Patterns of ecosystems	Patterns of ecological features and components	Individual ecological features and components
Character of information	Mostly qualitative	Qualitative and quantitative	Mostly quantitative

acres, whereas a regional forester implementing a strategy to conserve an endangered predator may need to consider a scale of several million acres. The following general guidelines can help determine which scale is appropriate for the job.

1. *Develop explicit objectives.* Too often, inappropriate scales are used to answer important questions because management and planning objectives are not clearly stated. The objectives must be clear enough so that the relevant ecological (or social) processes, driving variables, and the scale of analysis are obvious.

2. *Identify the components of the analysis.* For example, a wildlife habitat analysis is likely to include various components for a variety of species (i.e. nest sites, habitat patterns, feeding areas, amount of edge, patch size, and connectivity). The appropriate scale for assessing wildlife habitat will vary by species and habitat components. Environmental impact statements are especially sensitive to the scale of analysis because they frequently involve assessing multiple impacts and require examination of alternative actions. One scale of analysis will rarely be adequate. Different scales, determined by the types of potential impacts and details of the alternatives, will be needed.

3. *Planning analysis should encompass the space and time over which management actions could operate.* At the start of any analysis, it's worth asking what might happen at both smaller and larger or shorter and longer scales as a result of a management action. Coarser scales provide context while finer scales help identify mechanisms that create impacts. Unexpected influences can originate from events at larger and smaller scales. For example, large scale or extended drought events may exacerbate the impact of a modest water diversion that appears to have minimal impact in years of normal precipitation in the local area. Short-term disturbances in many ecosystems may appear as background "noise" over the longer time periods, while small, persistent, cumulative effects that individually appear insignificant in the short term can have major long-term consequences. One should examine the processes and phenomena operating at larger and smaller scales as sources of surprise or potential influences within the domain of the planning landscape. Having data from larger and smaller scales can help insure future information needs. This practice should be especially encouraged where sampling at additional scales can be done at little extra cost. Digital technologies now allow managers to create spatial data tools to make comparison over a range of scales.

4. *Use an ecosystem approach as a unifying tool to identify meaningful linkages among different scale levels and different categories of hierarchy* (physical systems, biological, social, and organizational hierarchies).

5. *Focus the analysis and limit unnecessary or confusing detail* by identifying up front the constraints that larger and finer scale processes, mechanisms, and decisions impose.

SELECTED READINGS

Allen, T.F.H. and T.W. Hoekstra. 1992. *Toward a Unified Ecology.* Columbia University Press, New York.

Bailey, R.G. 1996. *Ecosystem Geography*. Springer-Verlag, New York.

Gunderson, L., C.S. Holling and S. Light. 1995. *Barriers and Bridges to the Renewal of Ecosystems and Institutions*. Columbia University Press, New York.

Turner, M.G. and R.H. Gardner. 1991. Quantitative Methods in Landscape Ecology: An Introduction. In: M.G. Turner and R.H. Gardner (eds.), *Quantitative Methods in Landscape Ecology: The Analysis and Interpretation of Landscape Heterogeneity*. New York, Springer-Verlag.

Ecological Classification

Why Ecological Classification Is Important To Ecological Stewardship

Anyone traveling over a high mountain pass sees marked changes in the landscape as one climbs from the valley to the crest and back down the other side. These visible changes reflect variations in climate, hydrology, geology, soils, and species composition over the landscape. The life zones developed by von Humboldt and other nineteenth century geographers were early forms of ecological classification based on obvious changes in flora and fauna over altitudinal and latitudinal gradients (Fig. 1). Our understanding of ecosystems and their variation across space has come a long way since then. Today, ecological classification is a scientifically rigorous approach to systematically organize many kinds of information about ecosystems. Much of this information is less obvious to the casual observer than the variations noted by earlier geographers, but is critically important to understanding similarities and differences between ecosystems. In short, ecological classification provides a spatially explicit framework to organize knowledge about the biophysical components of the environment and their interactions over space and time.

This summary was drafted by Nels Johnson with contributions from Connie Carpenter, Dennis Grossman, Mark Jensen, and Paul Cunningham. It is based on the following chapters in Ecological Stewardship: A Common Reference for Ecosystem Management, Vol. II:

Constance A. Carpenter, Wolf-Dieter N. Busch, David T. Cleland, Juan Gallegos, Rick Harris, Ray Holm, Chris Topik, and Al Williamson. "The Use of Ecological Classification in Resource Management."

Dennis H. Grossman, Patrick Bourgeron, Wolf-Dieter N. Busch, David Cleland William Platts, G. Carleton Ray, C. Richard Robins, and Gary Roloff. "Principles for Ecological Classification."

KEYWORDS: ecotype, ecoregion, maps, information framework, aquatic ecosystems, terrestrial ecosystems

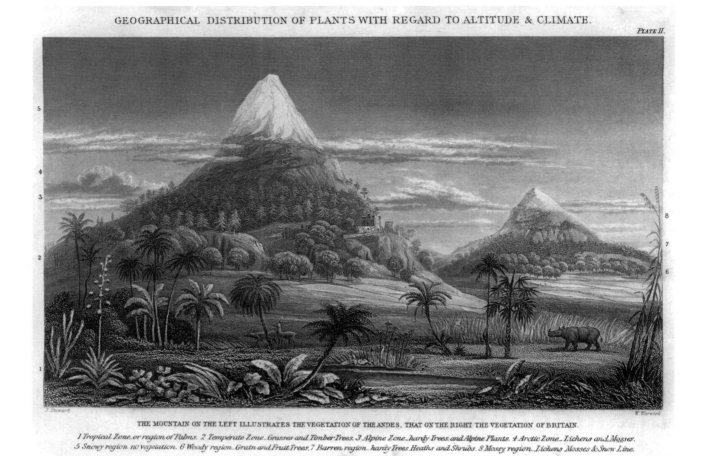

GEOGRAPHICAL DISTRIBUTION OF PLANTS WITH REGARD TO ALTITUDE & CLIMATE.

PLATE II.

THE MOUNTAIN ON THE LEFT ILLUSTRATES THE VEGETATION OF THE ANDES, THAT ON THE RIGHT THE VEGETATION OF BRITAIN.

1 Tropical Zone, or region of Palms. 2 Temperate Zone, Grasses and Timber Trees. 3 Alpine Zone, hardy Trees and Alpine Plants. 4 Arctic Zone, Lichens and Mosses. 5 Snowy region, no vegetation. 6 Woody region, Grain and Fruit Trees. 7 Barren region, hardy Trees Heaths and Shrubs. 8 Mossy region, Lichens Mosses & Snow Line.

Fig. 1. Early forms of ecological classification developed by nineteenth century geographers were based on obvious changes in flora and fauna over altitudinal and latitudinal gradients.

Ecological classification is important to ecological stewardship for several reasons. First, ecological classification is an indispensable information organizing tool. At any particular place and time, an ecosystem is composed of a potentially bewildering array of biophysical variables. Classification helps researchers and managers group these variables into unique combinations and distinguish one area from another on the basis of differences in composition, structure, and processes. Second, ecological classification enables mappers to represent biophysical factors, such as disturbance regimes and hydrologic processes, that are critical to resource management planning. Maps that simply represent visible landform and vegetation have limited potential as a planning tool. Behind every ecological classification map is an enormous set of information that managers and researchers can use for a variety of objectives. For example, ecological classification can help develop integrated resource inventories, model the cumulative impacts of management decisions, organize regional and landscape planning assessments, and design monitoring systems.

KEY FINDINGS

Ecological classification provides a foundation for many research and management activities in ecosystem management. Successful approaches to ecological classification require a thorough understanding of basic ecological principles and the management objectives of the users.

> **Box 1**
> **Definition of Key Terms**
>
> *Classification*: The systematic ordering of information such that there is more similarity between samples within a type than between different types.
> *Ecological Classification*: The grouping of similar items to provide a framework for organizing our knowledge about ecosystems. This grouping can include biological elements and variables (e.g., species distribution, species richness, plant community structure, habitat area and shape) and physical features and processes (e.g., soil type, precipitation, fire frequency, topography), or some combination of biotic and abiotic features.
> *Ecosystem Classification*: Basic units on the ground that represent biological components associated with the ecological patterns and processes that structure their association.
> *Ecological Map Unit*: A spatial polygon (land or aquatic unit) that has a predictable combination of ecosystem components (i.e. composition, structure, and function).

✓ **Ecological classification uses rules and class definitions to group biological and environmental variables into sets of observational units** (Box 1). These units can be defined at various scales. For example, ecological classification may be used to classify small areas of several hundred hectares (ecotypes), larger areas with recurring patterns of ecotypes over hundreds of square kilometers (ecological units), or very large areas, perhaps a hundred thousand square kilometers or more, that have groups of ecological units that function in a similar way (ecoregions).

✓ **Ecological classification should be designed with a clear purpose or management objective in mind.** There is no one ecological classification system that will work for all purposes. Rather, a classification system should be tailored to the management or research objective(s) it is being used for. For example, if the objective is to establish a monitoring network for aquatic ecosystems, a terrestrially-based ecological classification will not be appropriate (see below). To ensure that an ecological classification is suitable for management objectives, classification scientists should work closely with resource managers from early design, through field testing and final revisions to the system. Classification systems that don't jibe with managers' observations and experiences in the field are unlikely to be used.

✓ **Ecological classification is an iterative process involving field observation, statistics, and numerical modeling.** At fine spatial scales, ecological units are often identified by direct observation or sampling of flora, soils, physiography, and other variables in the field. At coarser scales, aerial photographs, satellite imagery, and previously mapped and ground-truthed data may be used to define ecological units. After an initial classification is made, units are often refined through additional field sampling and statistical analysis. Once large data sets are on hand, a variety of multivariate methods may be employed to identify patterns and to reduce the number of variables. For example, multivariate methods might be used to detect patterns in forest canopy cover, soil texture classes, or plankton distribution. If possible, such patterns should be verified through field observation or by comparison with other reliable data sets.

✓ **Aquatic ecosystems require distinct classification approaches.** Aquatic ecosystems are highly dynamic compared to terrestrial systems. Physical and biological variations in aquatic ecosystems are controlled by different factors than those that control important features of adjacent terrestrial ecosystems. Aquatic classification systems, therefore, should not be assumed to nest easily, if at all, within terrestrial classification systems. Aquatic systems are generally classified by analysis of physical features, such as temperature, salinity, watershed topography, the dendritic structures of river systems, estuarine structure, and the like. This is in strong contrast to the more biotic approach of land systems. Obviously, the need to use different classification approaches in terrestrial and aquatic ecosystems poses a dilemma for managers seeking integrated approaches to resource management. Using separate classification schemes, however, is better than omitting aquatic systems or assuming they are reasonably represented in terrestrial ecosystem classifications.

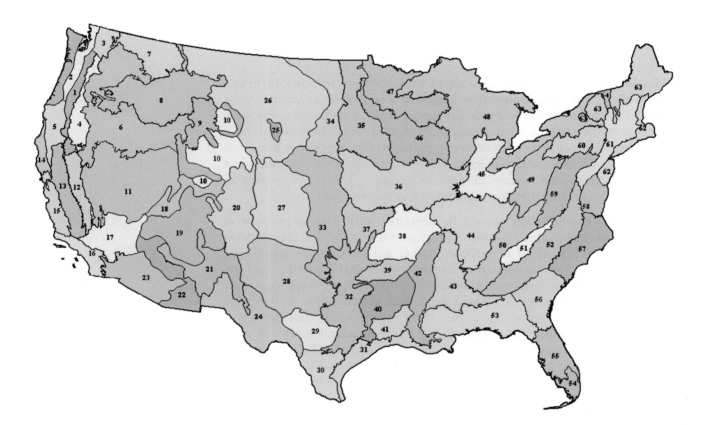

Fig. 2. The Nature Conservancy's ecoregional classification of the United States. (Courtesy of The Nature Conservancy.)
1 = West Cascades and Coastal Forests. 2 = Puget Trough and Willamette Valley. 3 = North Cascades. 4 = Modoc Plateau and East Cascades. 5 = Klamath Mountains. 6 = Columbia Plateau. 7 = Canadian Rocky Mountains. 8 = Idaho Batholith. 9 = Utah Wyoming Rock Mountains. 10 = Wyoming Basins. 11 = Great Basin. 12 = Sierra Nevada. 13 = Great Central Valley. 14 = California North Coast. 15 = California Central Coast. 16 = California South Coast. 17 = Mohave Desert. 18 = Utah High Plateau. 19 = Colorado Plateau. 20 = Colorado Rock Mountains. 21 = Arizona–New Mexico Mountains. 22 = Apache Highlands. 23 = Sonora Desert. 24 = Chihuahuan Desert. 25 = Black Hills. 26 = Northern Great Plains Steppe. 27 = Central Shortgrass Prairie. 28 = Southern Shortgrass Prairie. 29 = Edwards Plateau. 30 = Tamaulipan Thornscrub. 31 = Gulf Coast Prairies and Marshes. 32 = Crosstimbers and Southern Tallgrass Prairie. 33 = Central Mixed-Grass Prairie. 34 = Northern Mixed-Grass Prairie. 35 = Northern Tallgrass Prairie. 36 = Central Tallgrass Prairie. 37 = Osage Plains/Flint Hills Prairie. 38 = Ozarks. 39 = Ouachita Mountains. 40 = Upper West Gulf Coastal Plain. 41 = West Gulf Coastal Plain. 42 = Mississippi River Alluvial Plain. 43 = Upper East Gulf Coastal Plain. 44 = Interior Low Plateau. 45 = North Central Tillplain. 46 = Prairie-Forest Border. 47 = Superior Mixed Forest. 48 = Great Lakes. 49 = Western Allegheny Plateau. 50 = Cumberlands and Southern Ridge and Valley. 51 = Southern Blue Ridge. 52 = Piedmont. 53 = East Gulf Coastal Plain. 54 = Tropical Florida. 55 = Florida Peninsular. 56 = South Atlantic Coastal Plain. 57 = Mid-Atlantic Coastal Plain. 58 = Chesapeake Bay Lowlands. 59 = Central Appalachian Forest. 60 = High Allegheny Plateau. 61 = Lower New England/Northern Piedmont. 62 = North Atlantic Coast. 63 = Northern Appalacnian/Boreal Forest. 64 = St. Lawrence Plain.

✓ **The relationship between classification and mapping is often confusing for scientists and managers alike.** In its purest form, ecological classification is independent of mapping. It is a scientific process of methodically arranging units of quantitative information into classes or groups that possess common properties. A thoroughly developed ecological classification may exist only in a tabular or digital database and may not be mapped at all. Mapping, however, is a powerful way to represent the spatial distribution of ecological units, and is a particularly useful tool for various planning and management objectives (Fig. 3). Still, maps usually omit many variables and relationships that are described in an ecological classification. Researchers and managers may need to go to the data that is behind the maps to take full advantage of the information contained in an ecological classification.

✓ **More integrated approaches to ecological classification are needed.** Most ecological classifications have been narrowly defined to represent only specific taxa, landform, vegetation, climate, or geography. This fragmented approach to ecological classification hinders integrated approaches to resource planning and management over large geographic areas. To be useful in ecosystem management, greater integration of biological and physical variables is needed. Terrestrial and aquatic systems should both be classified in the context of landscape-level planning and management. And, if coastal systems are to be classified, they must be based on the interactions of land, sea, and atmosphere, which define the coastal zone. Advances in several areas of ecological theory will help improve the usefulness of ecological classification. In particular, research on landscape pattern recognition, ecosystem dynamics, biological and environmental thresholds, and hierarchy theory will generate improved techniques for integrated approaches to ecological classification.

PRACTICAL CONSIDERATIONS

Ecological classification can be useful for a range of management activities and objectives. In particular, ecological classification can play important roles in resource management, resource inventory and monitoring, biodiversity conservation, and landscape-level planning and assessment (Box 2).

Ecological classification can help to structure integrated resource inventories. Ecological classification is a useful tool in conducting resource inventories for several reasons. First, a classification system provides a framework to develop a stratified approach to resource sampling. That is, sampling can be designed to be representative of ecotypes or ecological units found at the watershed, landscape, or regional level. Second, ecological classifications allow researchers and planners to correlate a number of biological and environmental factors with the distribution and abundance of resources being inventoried. This can enable planners to predict where resources of interest are most likely to be found. Third, ecological classification and resource inventories can be synergistic. Field sampling for either can strengthen the other. If properly designed, resource management agencies may be able to gain efficiencies by planning field sampling to collect information that is simultaneously useful to ecological classification and resource inventories.

Ecological classification can help to establish reference conditions. Ecological classification is usually based on some definition of "natural conditions." Therefore,

---

Box 2
Ecological Classification in Natural Resource Planning and Management

Resource Inventory. Resource inventory applications include:
• Design of stratified resource sampling methods;
• Ecosystem basis for natural resource information collection, and;
• Ecosystem basis for reporting and mapping of natural resources.

Resource Management. Resource management applications include:
• Evaluation of sites for roads, power lines, buildings, and other infrastructure;
• Ecologically based management prescriptions for forestry, wildlife, fisheries, recreation, agriculture, erosion control, minerals and water, and;
• Definition of desired future conditions.

Biodiversity Conservation. Biodiversity conservation applications include:
• Design of biodiversity inventory and sampling;
• Identification of priority conservation sites;
• Design of ecologically representative protected area networks;
• Habitat suitability analysis, and;
• Design of ecological restoration projects.

Landscape Planning and Assessment. Planning and assessment applications include:
• Environmental impact assessments;
• Regulatory screening of proposed developments in wetland areas;
• Ecosystem and landscape monitoring;
• Study of pattern and process in relation to scale, and;
Predictive modeling of management mpacts.

---

the definition of an ecotype or an ecological unit can be useful in establishing reference conditions for various purposes. These might include developing a definition of "desired future conditions" for management goals, assessing environmental change, or gauging the success of restoration activities. Ecological classifications can also be a useful tool for establishing standards and criteria in monitoring systems designed to measure environmental quality and ecosystem integrity.

Ecological classification can help identify conservation priorities. Ecological classification is indispensable for assessing the representativeness of conservation areas. The starting point for many efforts to identify conservation priorities is ecological classification. For example, the "gap analysis" program sponsored by the U.S. Fish and Wildlife Service used ecological classification to identify ecosystems that were poorly represented in existing conservation areas. Likewise, The Nature Conservancy and other conservation groups have developed detailed classification systems to assist them in targeting limited financial and human resources.

Ecological classification has become an important tool in aquatic ecosystem management. Ecological classification has assisted aquatic ecosystem management in several areas, including water quality assessment, fish productivity modeling, fish habitat modeling, and assessment of adjacent land management impacts. For example, the National Water Quality Assessment Program (NAWQA) surveys the status and trends of ground and surface water quality in the United States. This U.S. Geological Survey (USGS) program collects physical, chemical, and biological data for watersheds and aquifers, in part, according to aquatic ecosystem classification units. Wetland ecosystem classification has also become an important and sometimes controversial tool in land-use planning and regulation. The U.S. Army Corps of Engineers, the U.S. Environmental Protection Agency, and state and local planning agencies use wetland classifications to carry out their regulatory and planning responsibilities.

Ecological classification can provide a useful tool for determining site potential for agriculture and forestry. Common applications of terrestrial vegetation and site classifications include helping to assess potential site productivity, determining appropriate management practices, and identifying suitable crop and tree species. Typically, such classifications emphasize the use of climatic, soil, and vegetation data. In the case of forestry, ecological classification can help to predict successional processes and characterize disturbance regimes. Such information can help forest managers design silvicultural systems that mimic natural ecosystem processes, thereby reducing environmental risks and improving conditions for biodiversity.

SELECTED READING

Bailey, R.G., M.E. Jensen, D.T. Cleland, and P.S. Bourgeron. 1993. Design and Use of Ecological Mapping Units. In: M.E. Jensen, and P.S. Bourgeron (eds.), *Eastside Forest Ecosystem Health Assessment.* Vol. 2. USDA, Portland, OR.

Busch, W.-D.N., and P.G. Sly (eds.). 1992. *The Development of an Aquatic Habitat Classification System for Lakes.* CRC Press, Boca Raton, FL.

Cowardin, L.M., V. Carter, F.C. Golet, and E.T. LaRoe. 1979. *Classification of the Wetlands and Deepwater Habitats of the United States.* U.S. Fish and Wildlife Service, Washington, DC.

ECOMAP [Ecological Classification and Mapping Task Team]. 1993. National Hierarchical Framework of Ecological Units. Unpublished administrative paper. USDA Forest Service, Washington, DC.

Environment Canada. 1989. Ecoclimatic Regions of Canada, first approximation (with map at a scale of 1:7,500,000). *Ecological Land Classification Series* No. 23. Environment Canada, Ottawa, Canada.

Frissell, C.A., W.J. Liss, C.E. Warren, and M.D. Hurley. 1986. A hierarchical framework for stream classification: Viewing streams in a watershed context. *Environmental Management* 10: 199–214.

Grossman, D.H., D. Faber-Langendoen, A.S. Weakley, M. Anderson, P. Bourgeron, R. Crawford, K. Goodin, S. Landaal, K. Metzler, K. Patterson, M. Pyne, M. Reid, and L. Sneddon. 1998. International Classification of Ecological Communities: Terrestrial Vegetation Of The United States. *The National Vegetation Classification System: Development, Status, and Applications.* Vol. 1. The Nature Conservancy.

Hayden, B.P., G.C. Ray, and R. Dolan. 1984. Classification of coastal and marine environments. *Environmental Conservation* 11: 199–207.

Higgins, J., M. Lammert, D. Grossman, and M. Bryer. 1996. "A Classification Framework for Freshwater Communities: Proceedings of The Nature Conservancy's Aquatic Community Classification Workshop, April 9–11, 1996, New Haven, Missouri." The Nature Conservancy, Arlington, VA.

Wiken, E.B. (ed.), 1986. Terrestrial Ecozones of Canada. *Ecological Land Classification* No. 19. Environment Canada, Ontario, Canada.

Humans as Agents of Ecological Change

♦ *Human Roles in the Evolution of North American Ecosystems*

♦ *Cultural Heritage Management*

♦ *Producing and Using Natural Resources*

♦ *Ecosystem Sustainability and Condition*

♦ *Ecological Restoration*

♦ *Population and Tradeoffs in Land Use*

Human Roles in the Evolution of North American Ecosystems

Why Human Roles in the Evolution of Ecosystems Are Important To Ecological Stewardship

Agricultural fields, pine plantations, channelized rivers, cities, and suburban subdivisions are obvious examples of how human actions have shaped today's landscapes in North America. But even contemporary wildlands have often been influenced by human activities, many by Native American cultures that disappeared more than a century ago. For more than 12,000 years, a long history of human interaction has affected the structure and composition of most ecosystems to varying degrees. This environmental transformation has escalated dramatically as humans organized themselves first around hunting and gathering, then around agriculture, and eventually industrialization. What appears today as "natural" landscapes are often the product of centuries or millennia of environmental manipulation by Native Americans and, more recently, by European Americans and immigrants from around the world.

This summary was drafted by Nels Johnson and Richard D. Periman with contributions from Henry T. Lewis and Connie Reid. It is based on the following chapters in Ecological Stewardship: A Common Reference for Ecosystem Management, Vol. II:

Bonnicksen, T.M., M.K. Anderson, H. Lewis, and C.E. Kay. "Native American Influences on the Development of Forest Ecosystems."

Periman, R.D., C. Reid, M.K. Zweifel, G. McVicker, and D. Huff. "Human Influences on the Evolution of North American Landscapes: Methods and Applications for Managing Ecosystems."

KEYWORDS: Native Americans, European Americans, fire regimes, agriculture, culture

Box 1
"Cultural" Fire Regimes

If there was an overall point to human-set fires, it was to establish and then maintain significantly different conditions than those found to exist naturally. The maintenance of such conditions was not the simple replication or the intensification of natural fire regimes; to a very large measure it involved their replacement. Fires were set and managed for a variety of purposes: to control plant diseases and insect infestations; to increase the range and abundance of desirable plant species; to stimulate plant growth; to minimize the potential for severe wildfires; and to reduce undergrowth and facilitate hunting. The distinguishing features of cultural fire regimes include: (1) burning different kinds of environments in different seasons; (2) altering the frequencies with which fires are set and reset over varying periods of time; (3) manipulating the intensity of fires; (4) deliberately choosing sites where fires are set and alternately, where areas are to be protected from fire, and; (5) using a range of natural and artificial controls to limit the spread of human-set fires, such as times of day, winds, fuels, slope, relative humidities, natural fire breaks, etc. With this information in hand, prescribed burns can be designed to mimic the cultural fire regimes that helped shape forest and grassland ecosystems (Fig. 1).

Managers are more likely to make decisions that benefit ecosystem health and resource sustainability decisions when they understand the cumulative effects of human-environment interactions. For example, knowledge of the burning seasons and practices of Native Americans can be used to restore ecosystem health in a degrading system where frequent fires were the norm for centuries (Box 1). A manager who reintroduces the use of fire without understanding earlier burning practices, however, may actually contribute to

Fig. 1. Prescribed fire has become an important tool in ecosystem management. Information on how humans have used fire over time in an area can be used to design an appropriate prescribed fire management plan. (Photo courtesy of USDA Forest Service.)

further ecosystem degradation. Archaeologists, anthropologists, paleo-biologists, geologists, and environmental historians can help managers understand these details and their implications for contemporary management issues.

While today's society may have different goals than earlier occupants of the land, their knowledge, tools, and practices provide a valuable legacy for a range of ecosystem management objectives. In particular, an understanding of human roles is often essential in ecological restoration, and may be relevant in wildlife management, range management, and forestry. Even the recovery of endangered species may depend on knowing how humans have interacted with the species and its habitat over the centuries. The information and tools available to help decipher earlier human-environment interactions are growing even though our predecessors have long since departed. This knowledge can help today's citizens and natural resource managers shape more constructive human roles in ecosystem processes.

KEY FINDINGS

Systematic research on human roles in shaping the composition and structure of North American ecosystems is a relatively new field. In trying to understand these roles, it is important to recognize the following:

✓ **Native Americans were significant agents of environmental change for at least 12,000 years.** Through their use of fire and selective harvesting of plants and animals, North America's earliest settlers added to the natural forces of climate, geology, lightning, disease, and predation that shaped the continent's ecosystems. Prehistoric peoples also influenced ecosystems through activities such as: sowing and broadcasting of seeds; transplanting shrubs and small trees; diverting water for irrigation and erosion control; pruning and coppicing of plant communities; construction of water diversion structures for erosion control; and selective harvesting of wild plants.

✓ **As Native American cultures developed, more elaborate systems of resource management evolved.** In the Southwest, where agriculture was an important subsistence strategy, people constructed fields; harvested large amounts of wood for fuel and construction; and constructed earthworks and canals for homes and irrigation. In the Eastern Woodlands, agriculture and extensive trade networks supported a number of large population centers. On the West Coast, an abundance of natural resources supported a diverse array of cultures built around harvesting fish and marine life. When Europeans arrived, humans had already significantly influenced ecosystems across the continent (Box 2).

✓ **North American ecosystems were extensively altered after 1492.** For management purposes, the historic period in the development of North American ecosystems began in 1492 with the advent of European colonization in the New World. A 500-year succession of environmental changes on a scale not encountered since the end of the Ice Age followed the arrival of Spanish, Portuguese, English, and French explorers and settlers. This new era of human influence, chronicled in the journals and official documents of these early explorers, settlers, military, and church officials, is an important part of the 12,000-year continuum of human land use on this continent.

**Box 2
The Myth of the Empty Continent**

By 1492, Native Americans had modified ecosystems across the continent. In many areas they influenced the extent and composition of forests, created and expanded grasslands, influenced the mosaic structure of brush lands, and rearranged micro-relief via the construction of countless earthworks. The use of fire enabled even relatively small populations to have significant ecological impacts at a broad geographic scale. The landscapes encountered by the first European explorers and colonists had been systematically harvested for favored wildlife and plants, regularly set on fire by their human inhabitants for a variety of reasons, and were laced with villages, trails, and agricultural fields (Fig. 2).

Fig. 2. Native Americans developed intensive agriculture and large settlements on parts of the landscape, particularly in river valleys and along coastal areas in what is now the eastern United States. (Photo courtesy of the Library of Congress.)

Box 3
Shifting Uses of the Landscape

European trapping, mining, logging, and agricultural practices soon eclipsed the ecological impacts of the Native Americans they displaced. Growing populations, expanding trade, changing technologies, and local natural resource scarcity meant that many landscapes went through a rapid succession of human uses and practices (Fig. 3). Landscapes were transformed not just once, but often two, three, and four times within the span of a century or two. Each transformation — for example, from forest to patchy agriculture to hillsides stripped of trees to fuel ironmaking in a New England watershed — carried with it different implications for the ecological composition, structure, and function of the landscape. This history continues to influence the condition of many ecosystems today. In some areas, natural forests reinvade abandoned agricultural fields. In other areas, suburban development leapfrogs into forested and agricultural areas while tree plantations replace naturally reproducing forests.

✓ **Early European influences grew as a global economic system developed.** This economic system evolved from the expansion of European enterprises overseas between 1500 and 1800 and led to the rapid growth of world trade networks. Efforts to promote a favorable balance of trade and obtain colonies which could supply raw materials to feed the expanding industrial revolution helped to create the system of multinational trade, banking, and resource development that circles the globe today. Spanish, French, and English explorers competing for resources in the Americas claimed the land and resources they encountered. The western hemisphere was regarded as a storehouse of wealth to be exploited for financing wars and expanding European power and influence. The search for precious minerals and agricultural land stimulated this expansion throughout North America. Human ecological history needs to be understood in this dynamic historical context (Box 3).

✓ **A variety of methods are available to help decipher human environmental influences.** Human behavior and the nature of cultural change are intertwined with environmental change. Ecosystem research and analysis should reflect this dynamic process and include methods and procedures to collect and analyze information on human-environment interactions. Documenting the history of people and the environments in which they lived involves using a variety of information sources. These include ethnohistoric and ethnographic literature; museum study of cultural plant products; ethnobotany; archaeology; paleoecology; geology and pedology; pyrodendrochronology, historical photograph interpretation; experimentation and observation. Through application of these methods, an accurate range of variability, including human and climatic influences, can be understood, and a more complete scope of management options considered.

✓ **The methods chosen will depend on the area and the questions to be researched.** For example, research questions focused on the vegetational changes within a landscape that occurred subsequent to the policy of active wildfire suppression could use historic photographs dating prior to fire control. To cross-reference and substantiate the landscape conditions depicted in the photos and achieve a higher level of accuracy, additional sources of information, such as land surveys, written documents, soil profiles, and paleobotanical records (e.g, pollen records or tree rings), can be used. By approaching both the natural and cultural history of a landscape, with data derived from a wide variety of sciences and sources, we can synthesize an understanding of the human and natural processes that interacted in creating an ecosystem.

✓ **The study of humans as agents of ecological change for ecosystem management is relatively new**. Despite growing recognition in recent decades that humans have influenced ecosystems in North America for centuries and even millennia, the incorporation of information on long-term human environment interactions into natural resources management is a new endeavor. We are only beginning to understand the complexity of human effects on environmental systems through time. A great deal of research remains to develop a reasonably complete understanding of how humans have altered North American ecosystems over the centuries.

✓ **Spatial and temporal scales are important when considering how, when, and where people have changed the environment**. The scale at which human action becomes a major factor in environmental change has varied with time, space and culture. It is therefore necessary to examine factors of duration, intensity, and periodicity at specific temporal and spatial scales. The relationship between human economic systems and particular environments is relative: a certain type of economic activity, fur trading for example, may have dramatically different environmental effects, depending on the species and the ecosystem. A minute change in one environment

Fig. 3. New technologies introduced in the nineteenth century, such as railroads and sawmills, greatly expanded and accelerated human influences on the environment. (Photo courtesy of U.S. Library of Congress.)

could be perceived as a major change in another. When conducting ecosystem analysis, these aspects of scale are extremely important. A well defined scale of inquiry is crucial whether compiling data for a regional study or focusing on a single watershed. Ecosystem analysis need not include a study of human effects on the entire globe for the last million years, but it is important that appropriate judgement be used in establishing the temporal parameters of an investigation. One size does not fit all, and neither does any single methodology.

✓ **The role of humans in shaping ecosystems is sometimes distorted to support particular views**. Differing opinions concerning public land management are deeply rooted in values. Value-based debate may ignore facts, or interpret them from a narrow viewpoint. Historic records and other information have fallen victim to these circumstances. History may be used to support the argument that the development of natural resources has actually improved the environment. Interpretations drawn from the journals of early western explorers are sometimes used to support this argument. On the other side of the debate, ranchers, miners, and other pioneers are often characterized as wanton and greedy destroyers of ecosystems. A contextual approach to analyzing historical human effects on ecosystems shows the development of many of our contemporary values and perceptions concerning natural resource management.

PRACTICAL CONSIDERATIONS

Local circumstances will determine what steps and methodologies are most appropriate for unraveling the role of humans in the evolution of contemporary landscapes. The following steps, however, are likely to be useful in any setting.

Determine project goals and design appropriate research questions to direct and focus research efforts. The methods, approaches, and techniques, when used in combination with the appropriate research questions, produce a tangible record of human influence on the environment, and a wealth of knowledge about climate change and ecological processes. The methods chosen will depend on the area to be researched, and the questions one wishes to address.

Identify the temporal and spatial scales appropriate to the research questions. Appropriate scale of analysis is important for both scientific, management, and cost effectiveness. Too little understanding of the spatial extent or duration of human influences can result in faulty interpretations of ecosystem dynamics, processes, structure, and function. On the other hand, a study designed at too large a scale in either space or time can be very expensive and generate information that is not specific enough to help answer management questions.

Understanding the roles of humans in environmental change takes time This is also true of timber, range, soils, and biological analyses. While a review of existing information, agency records, and archival material may be appropriate for general studies of past human environmental influences, there are cases, usually at the landscape and project/site levels of analysis, when new field data are needed.

Use an interdisciplinary approach. Experts from the fields of anthropology, archaeology, history, paleoecology and paleontology, geology, and a wide variety of other disciplines can be consulted to provide the necessary skills, methods, and analyses for addressing this topic. Government agencies often have staff with training in these fields. Also, this expertise is available in colleges, universities, and the private sector.

Become familiar with case studies and example projects. A variety of studies of human and environmental interaction are ongoing on public and private lands throughout the nation. Becoming familiar with agency and academic archaeological, historical, paleobotanical, or other research projects in your state, region, or district will help provide insight into past human processes. Agency specialists may be collecting information crucial to management decision making right on your unit; see what they have to offer.

SELECTED READING

Blackburn, T.C., and K. Anderson (eds.). 1993. *Before the Wilderness: Environmental Management by Native Californians.* Ballena Press, Menlo Park, CA.

Betancourt, J.L., T.R.V. Devender, and P.S. Martin (eds.). 1990. *Packrat Middens: The Last 40,000 Years of Biotic Change.* University of Arizona Press, Tucson, AZ.

Bonnicksen, T.M. 1991. Managing biosocial systems. *Journal of Forestry* 89(10): 10–15.

Butzer, K.W. 1982. *Archaeology as Human Ecology: Methods and Theory for a Contextual Approach.* Cambridge University Press, Cambridge, MA.

Crumley, C.L. (ed.). 1994. *Historical Ecology: Culture, Knowledge, and Changing Landscapes.* School of American Research Press, Santa Fe, NM.

Gomez-Pompa, A. and A. Kaus. 1992. Taming the wilderness myth. *BioScience* 42(4): 271–279.

Gottesfeld, L.M. 1994. Aboriginal burning for vegetation management in northwest British Columbia. *Human Ecology* 22: 171–188.

Hammett, J.E. 1992. The shapes of adaptation: historical ecology of anthropogenic landscapes in the Southeastern United States. *Landscape Ecology* 7(2): 121–135.

Joukowsky, M. 1986. *A Complete Manual of Field Archaeology: Tools and Techniques of Field Work for Archaeologists.* Prentice-Hall, Englewood Cliffs, NJ.

Kay, C.E. 1995. Aboriginal overkill and native burning: Implications for modern ecosystem management. *Western Journal of Applied Forestry* 10: 121–126.

Lewis, H.T. 1978. Traditional uses of fire by Indians in northern Alberta. *Current Anthropology* 19: 401–402.

Pearsall, D.M. 1989. *Paleoethnobotany: A Handbook of Procedures.* Academic Press, Inc. San Diego, CA.

Pyne, S.J. 1982. *Fire in America: A Cultural History of Wildland and Rural Fire.* University Press, Princeton, NJ.

Reitz, E.J., L.A. Newsom, and S.J. Scudder (eds.). 1996. *Interdisciplinary Contributions to Archaeology: Case Studies in Environmental Archaeology.* Plenum Press, New York.

Cultural Heritage Management

Why Cultural Heritage Management Is Important To Ecological Stewardship

Heritage management — that is, the conservation and allocation of past and contemporary cultural phenomena — has traditionally been conducted outside the mainstream of natural resources management. Research and protection of heritage resources are often viewed as barriers rather than bridges to ecosystem management. Cultural remains and practices are typically seen as valuable only in social research, not in ecological analyses. This perspective is unfortunate since most ecosystems have been influenced by several millennia of environmental and cultural co-evolution. Resource managers who ignore historical processes deprive themselves of knowledge that can actually help rather than hinder their ability to reach ecological stewardship goals. In short, studying heritage resources provides a framework to identify baseline historic ecological conditions, detect trends and patterns of change, and to establish possible thresholds of ecosystem and cultural resiliency and sustainability (see Box 1).

This summary was drafted by Nels Johnson with contributions from Alan P. Sullivan, III, Joseph A. Tainter, Suzanne K. Fish, Clay Mathers, and Patricia Spoerl. It is based on the following chapter in Ecological Stewardship: A Common Reference For Ecosystem Management, Vol. II:

Sullivan, III, A.P., J.A. Tainter, and D.L. Hardesty. "Historical Science, Heritage Resources, and Ecosystem Management."

KEYWORDS: Historic variation, archeology, artifacts, Native Americans, European Americans

---

Box 1
Updating Cultural Heritage Management to Facilitate Ecological Stewardship

Traditionally, cultural heritage management has focused on large archaeological sites or "rich" concentrations of artifacts, such as the spectacular finds featured in *National Geographic* magazine. Although these remains are essential to understanding how earlier peoples occupied their perennial settlements, many clues about how humans have used the land are scattered across the landscape. Unfortunately, when dispersed remains are devalued or, worse yet, excluded from consideration in ecological stewardship planning, the usefulness of archaeological information in ecosystem management is diluted.

How can heritage management assume a more central role in ecosystem management? Some researchers advocate a new vision that seeks to integrate heritage management more fully into ecosystem approaches to natural resources management. Implementing this vision requires several steps. To begin, natural resource management agencies should explicitly recognize that ecosystems are products of a specific history that arose partly as a consequence of human activity. Next, they should emphasize the origins and extent of cultural landscapes, rather than focusing principally on large, rich archaeological sites. The nominating process for the National Register of Historic Places, for example, should incorporate the value of small archaeological sites scattered across the landscape — sites that provide essential clues to understanding past land-use systems. Such an understanding will make heritage management more relevant to ecosystem management needs, especially by helping to develop accurate reference conditions.

Knowledge of heritage resources is important to ecological stewardship in several ways. First, heritage resources are data banks that contain vital information to help reconstruct environmental conditions that prevailed at a given time. This information is essential in establishing accurate reference conditions for ecological restoration projects or assessing human impacts on ecosystems over time. To understand reference conditions, one needs to know how ecosystems have developed during 12,000 years or more in response to societies that ranged from low-density foragers to high-density village agriculturalists (Fig. 1). Second, information from heritage resources, together with ecological records (e.g., tree-rings, pollen), can help resource managers establish baseline

Fig. 1. Major archaeological finds, such as villages remains and petroglyphs, have traditionally been the focus of cultural heritage management.

conditions at some chosen point in the past. By analyzing environmental conditions and heritage resources with regard to baseline conditions, researchers and managers can have a better idea of how ecosystems have changed and to what extent humans have influenced ecosystem development. Third, heritage resources provide valuable insights into how earlier peoples, including both Native American and European American cultures, managed and used their environment. Finally, many landscapes contain features and artifacts that encode significant religious, historical, and symbolic values. Resource managers have an ethical responsibility — and a legal responsibility on federal lands — to protect these resources as part of their management mandate.

KEY FINDINGS

✓ **Human societies and ecosystems have significantly influenced each other.** The once-common notion that small numbers of Native Americans inhabited a primordial garden is giving way to a more realistic view. Native Americans, for example, manipulated vegetation to select for early successional stages and to increase the abundance of particular plant and animal species. It now seems clear that Europeans encountered ecosystems that had been evolving for millennia along with, and in response to, growing Native American populations.

✓ **To define reference conditions, range of natural (or historic) variation, and restoration goals, it is essential to know how humans interact with the environment.** A historical understanding of the human role in shaping ecosystems and their evolutionary processes can yield practical resource management benefits. For example, humans were a major, perhaps dominant, ecosystem component in New Mexico's Arroyo Cuervo region of the Rio Puerco for more than 6,000 years. Human use during this time influenced a wide variety of ecosystem structures and processes. The disappearance of these peoples and their practices meant that ecosystems had to adjust to new conditions and establish new tolerance ranges. Models of pre-European ecosystem structure that do not incorporate the effects of human activity will be, at best, incomplete and, at worst, misleading. Lamentably, researchers lack much fundamental knowledge of how North American ecosystems functioned before A.D. 1500.

✓ **Scale is critical in cultural heritage research and management.** Just as scale is important to understanding such ecological processes as fire, it is also vital for understanding how humans interact with the environment. Many heritage studies evaluate sites individually and in isolation. For heritage management to contribute to ecosystem management, the appropriate scale of analysis may not be sites but larger spatial entities, such as cultural landscapes (see Box 2). Current heritage programs typically involve unintensive surveys (i.e., large distances between surveyors), little or no shovel testing, site definitions with high thresholds, and piecemeal evaluation of sites based mainly on "salient" characteristics (that is, large noticeable features). Yet, a program of heritage management conducted in this way may actually suppress data that are relevant to ecosystem management. To ensure that the full range of heritage resources is available for ecosystem management purposes, a heritage management program should

> **Box 2**
> **Cultural Landscapes**
>
> A first step in understanding how humans influenced past ecosystem processes is to delineate the landforms and vegetation zones used at various times. *Cultural landscape* is a concept that encompasses the biological and geophysical features of an environment as perceived, used, and modified by humans. For example, cultures in the middle Rio Grande basin in New Mexico during the Archaic period (5500 B.C. to A.D. 500) were predominantly hunter–gatherer societies. They lived in upper catchments and side canyons with reliable springs and topographic diversity that provided habitats for a variety of wild plant and animal species, upon which humans subsisted. The cultural landscape in the same region during the Pueblo IV period (A.D. 1300–1600) was strikingly different. People in this period abandoned and essentially did not use the upper watersheds and side canyons; instead, they practiced intensive agriculture in the Rio Grande valley bottom and built aggregated settlements on an unprecedented scale. Until the arrival of large-scale cattle ranching and irrigation in the late 19th century, this shift represented the most profound transformation of land and resource use during 12,000 years of human occupation.

include intensive survey (i.e., 15 meters or less between surveyors), artificial exposures (e.g., shovel testing) in areas of low visibility, site definitions with low thresholds, and the investigation of cultural landscapes, rather than individual sites.

✓ **Minimal legal compliance is an inadequate strategy for heritage management in the context of ecological stewardship.** Research and conservation of prehistoric and historic cultural resources on public lands are often triggered by legal provisions of the National Historic Preservation Act (1966), the Archeological Resources Protection Act (1979, as amended), and 36 CFR §800. These laws have contributed enormously to preserving our cultural heritage in the United States. Still, their bias towards protecting large sites and dense concentrations of artifacts frequently means that the interpretive and scientific potential of low-density sites and widely scattered artifacts may be overlooked (Fig. 2). To support ecosystem management, compliance with legal provisions should be conducted so that salient sites, low-density remains, and isolated artifacts are evaluated jointly, rather than individually, as evidence of past land-use systems. Future revisions to the Archaeological Resources Protection Act or 36 CFR §800, or special arrangements to implement them (such as programmatic memoranda), should be designed with such principles in mind.

PRACTICAL CONSIDERATIONS

Until recently, archeologists have tended to study features and sites that one could trip over, the assumption being that the only uncontaminated reliable data about the cultural past were from large, deeply stratified sites. While such sites will always have high value for some problems, these preferences may be excluding many kinds of heritage resources that could contribute to implementing an ecosystem approach. Researchers and resource managers can take a number of steps to expand the scope of cultural heritage management so that it generates more useful information for ecological stewardship.

Expand discovery protocols to increase the likelihood of finding low-density surface remains and isolated artifacts. Discovery strategies refer to such matters as survey

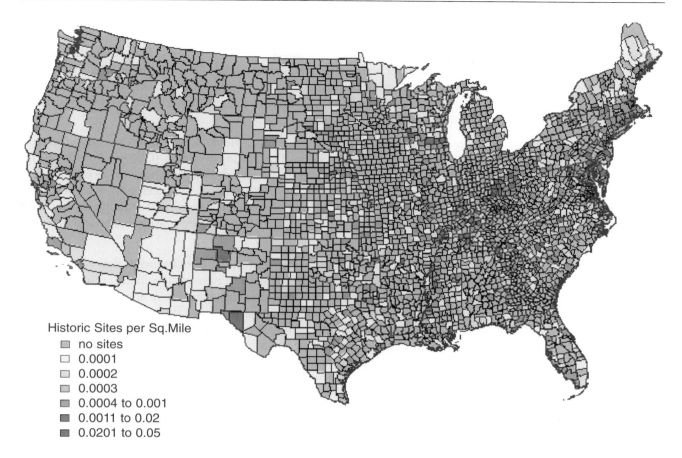

Historic Sites per Sq.Mile
- ☐ no sites
- ☐ 0.0001
- ☐ 0.0002
- ☐ 0.0003
- ■ 0.0004 to 0.001
- ■ 0.0011 to 0.02
- ■ 0.0201 to 0.05

Fig. 2. This map of archaeological sites on the National Register of Historic Places shows the density of sites per county. Funding and research tend to be biased toward those places with a high density of sites. This bias overlooks widely scattered sties that may yield vital clues to historic and prehistoric and uses. (Courtesy of National Park Service.)

intensity (i.e., inter-surveyor spacing), artificial exposure intensity (e.g., shovel-test spacing), and site definition. The standard for archaeological survey crew spacing in the USDA Forest Service Southwestern Region is 15–25 meters. Similar standards, which are followed in many parts of the United States, are actually an improvement over older methods of site discovery. Using more intensive survey techniques can yield considerably more information. For example, on the Hoosier National Forest in Indiana one contractor used 4.5 times more effort to survey an area half the size of an adjacent area surveyed by another contractor. The survey intensity paid off, however, because the first contractor found 35 times more archaeological sites per acre. The lesson here is that discovery techniques have a major influence on ascertaining the extent of the cultural heritage resource record.

Use low artifact-density thresholds to define what constitutes an archaeological site. Many site definitions involve artifact-density thresholds, such as five artifacts per square meter. If the surface properties of archaeological remains do not meet or exceed the thresholds, then there is a strong likelihood that those archaeological manifestations will not be incorporated into computer-based distributional studies. When several hundred or thousand such remains are excluded from regional data bases, models of prehistoric land use are inexact. The problem for heritage management is that the cutoff for calling an archaeological manifestation a "site" may be unjustifiably high, thereby excluding low-density surface scatters and isolated artifacts that often are important clues to land-use histories. This problem is relevant for both prehistoric and historic archaeology. To increase the value of cultural heritage information for ecosystem management, one should enter low-density surface remains and isolated artifacts into data bases for distributional analyses. In an ecosystem framework, site size and debris density are not the only indicators of significance.

Use advanced technologies and new strategies to assess larger areas for cultural heritage remains and significance. Global Positioning Systems (GPS), remote sensing, Geographic Information Systems (GIS), and laser ranging provide increasingly effective and inexpensive tools for heritage management. Another cost-effective strategy is to train biologists, foresters, and other field-oriented personnel to collect heritage information as they map habitats and conduct vegetation surveys.

Dismantle some resources to provide confirming or independent data regarding the human use of ecosystems. Archaeological and non-archaeological resources (e.g., old trees, bogs, packrat middens, alluvial exposures) are prime sources of paleo-environmental samples, such as macrobotanical remains, pollen, and faunal remains. The content of these resources is crucial for understanding an ecosystem's history and, hence, for framing the decisions that managers make regarding resource protection, allocation, and investigation. With ecosystem management, such an approach expands what it means to manage resources, because it entails a shift in orientation from find and protect, to find, protect and, when appropriate, consume heritage resources.

SELECTED READING

Balee, W. 1998. *Advances in Historical Ecology.* Columbia University Press, New York, NY.

Barker, J.P. 1996. Archaeological contributions to ecosystem management. *SAA Bulletin* 14: 18–21.

Butzer, K.W. 1982. *Archaeology as Human Ecology.* Cambridge University Press, Cambridge, UK.

Chambers, F.M. (ed.). 1993. *Climate Change and Human Impact on the Landscape.* Chapman and Hall, London, UK.

Crumley, C. (ed.). 1994. *Historical Ecology.* School of American Research Press, Santa Fe, NM.

Ebert, J.I. 1992. *Distributional Archaeology.* University of New Mexico Press, Albuquerque, NM.

Ellen, R.F. 1982. *Environment, Subsistence, and System: The Ecology of Small-Scale Social Formations.* Cambridge University Press, Cambridge, UK.

Lipe, W.D. 1995. The archeology of ecology: Taking the long view. *Federal Archeology.* (Spring): 8–13.

Mellars, P. 1976. Fire ecology, animal populations, and man: a study of some ecological relationships in prehistory. *Proceedings of the Prehistoric Society* 42: 15–45.

Moran, E.F. (ed.). 1990. *The Ecosystem Approach in Anthropology.* University of Michigan Press, Ann Arbor, MI.

Sullivan, A.P., III (ed.). 1998. *Surface Archaeology.* University of New Mexico Press, Albuquerque, NM.

Tainter, J.A. 1998. Surface archaeology: perceptions, values, and potential, pp. 169–179. In: A.P. Sullivan, III (ed.), *Surface Archaeology.* University of New Mexico Press, Albuquerque, NM.

Producing and Using Natural Resources

Why Natural Resources Production Is Important To Ecological Stewardship

Good science and innovative resource management practices are hallmarks of effective ecosystem approaches. But experience — particularly in many rural areas of the West — indicates the real test is whether ecosystem management can help reduce polarization over management objectives on public lands, where ecosystem management efforts are concentrated. Successfully fitting resource production outputs into the process of ecosystem management will both advance ecological stewardship goals and reduce confusion and controversy over public lands management.

This summary was drafted by Nels Johnson with contributions from Douglas MacCleery, Robert Ewing, and Mark Lorenzo. It is based on the following chapters in Ecological Stewardship: A Common Reference For Ecosystem Management, Vol. II:

MacCleery, D., and D.C. Le Master. "The Historical Foundation and the Evolving Context for Natural Resource Management on Federal Lands."

Johnson, M., J. Barbour, D. Green, D. MacCleery, S. Willits, M. Znerold, J. Bliss, S.L. Chiang, and D. Toweill. "Ecosystem Management and the Use of Natural Resources."

KEYWORDS: Natural resource management, forestry, fisheries, agriculture, public lands, multiple-use

Natural resource production activities are a minor or overlooked component of many federal ecosystem management projects. Understanding the relationship between natural resource production and ecosystem management goals, however, is fundamentally important for several reasons. First, some natural resources are extracted from most parts of the landscape — be it by farming, forestry, grazing, fishing, hunting, or mining extraction. Managing that larger landscape simultaneously for the protection of biodiversity and ecosystem values such as clean water is therefore a central challenge for many ecosystem management projects.

Second, many natural resource production activities can be designed to mimic or accommodate natural processes that are vital to ecosystem health. Such activities — silvicultural treatments or design of grazing allotments, for example — can play an important role in meeting ecological stewardship goals. At the same time, activities such as prescribed burning can reintroduce or reinvigorate natural ecosystem processes on resource production lands where they have disappeared.

Third, even as ecosystem approaches signal less intensive natural resource production on many parts of the federally managed landscape, human populations and resource consumption continue to grow. This means that reducing timber production in the Pacific Northwest is likely to result in increased harvests and adverse ecological impacts elsewhere, perhaps in more critical ecosystems. Therefore, it is important for resource managers, policy makers, and the public to recognize the tradeoffs at a variety of scales that may accompany decisions to use ecosystem approaches in natural resources management.

KEY FINDINGS

The paradigm for natural resource management on federal lands since the late 1950s has been the concept of "multiple-use management," or the simultaneous management of the larger federal landscape for several "outputs" such as timber, forage, and recreation. Much of the controversy over "ecosystem management" in the United States stems from confusion over whether ecosystem approaches represent a minor shift or a major change from such approaches as multiple-use and sustained yield management (see Box 1).

To understand the implications of ecosystem management for resource production, it is useful to compare goals between ecosystem management and multiple-use management. Table 1 compares biological, socio-economic, and management goals. Not surprisingly they both differ and overlap. Where the goals differ, they reflect broader changes in science and society. Similar changes in goals would likely be seen in education, health care, and urban planning.

Goals, or course, are easier to state than they are to accomplish. Achieving the ecosystem management goals will depend on how we choose to produce and consume natural resources. A variety of social, economic, and technological forces will help determine whether society can simultaneously maintain or improve ecological conditions and produce the natural resources it needs. Public behavior will likely be more important than choices made by natural resource managers in determining the success of ecosystem management for the following reasons:

Box 1
Ecosystem Management: Revolution or Evolution?

Is ecosystem management a modest evolutionary step from earlier multiple-use approaches, or is it a revolutionary departure? Certainly ecosystem management differs in two major respects: (1) it expands management objectives from a relatively few tangible resource outputs to a broader spectrum of values, uses, and environmental services, and (2) it requires consideration of social, biological, and economic interactions at a variety of spatial scales at local, regional, national, and international scales over time. However, like multiple use-sustained yield management, ecosystem management is a means to an end, rather than an end in itself. Providing human needs, including the production of natural resource commodities, are legitimate ends for which ecosystem management can provide useful tools.

History also suggests the answer is an evolutionary step — although a substantial one. For example, even before the official adoption of ecosystem management by most federal land management agencies in the early 1990s, many of its basic concepts had been developed and implemented to varying degrees. For example, a ground breaking 1979 publication edited by Jack Ward Thomas provided forest managers an insightful and systematic approach for integrating wildlife management objectives and timber management activities on large ownerships. Interagency task forces on the grizzly bear, the Greater Yellowstone Ecosystem, and other issues followed during the 1980s. These efforts brought forward new scientific concepts and a recognition that some species and ecosystem processes could only be maintained through coordinated actions at a broad geographic scale. By the early 1990s, growing conflicts over the environmental impacts of federal resource production activities and increased court involvement in settling disputes paved the way for administrative policies emphasizing ecosystem approaches on federal lands.

Table 1. Comparison of Goals: Ecosystem Management and Multiple Use–Sustained Yield Management

Ecosystem Management	Multiple Use–Sustained Yield Management
1. Biological	*1. Biological*
Biological diversity	Maintain habitats for featured species
Ecosystem and resource sustainability	Sustainability of renewable resources, i.e. achievement and maintenance in perpetuity of high-level outputs of renewable resources without impairment of the productivity of the land
Ecosystem health or integrity	Resource productivity
Production of resource outputs that "best meet the needs of the American people"	Production of resource outputs that "best meet the needs of the American people"
Integrated management	Individual or multi-resource management
2. Socio-economic	*2. Socio-economic*
Extension of internal agency resources through public and private partnerships (collaboration) and interagency cooperation	No directly comparable goal but partnerships used as means to achieve programmatic objectives
Social responsiveness	Community stability.
Political acceptability	Political acceptability
3. Resource Management	*3. Resource Management*
Risk minimization or aversion	Nothing comparable

✓ **Consumption of natural resources must be aligned with the capacity of natural systems to provide them**. Population, economic growth, and the policies of governments have already driven demand for some resources — fisheries, for example — beyond sustainable limits. Public policy will increasingly need to be directed toward creating positive incentives for resource conservation and more efficient use of natural and human resources. Recycling is an important element of a resource conservation policy. The fair market value policy for private use and consumption of natural resources should be more fully implemented, and the prices for federally owned resources should, in general, cover their marginal cost of production including any associated external environmental and economic costs.

✓ **Technology can be environmentally destructive, benign, or restorative: The choice is not whether technology, but how to manage it**. In the context of ecosystem management, this means to replace technologies that are environmentally destructive, to use natural resources more efficiently, to use natural resources that are abundant as substitutes for those that are scarce, and to extend the life of products.

✓ **Fair market value in the sale of natural resources from federal lands is a basic condition for sustainable natural resource production**. This policy has been articulated many times by Congress, although exceptions are widespread in practice. Exceptions should be allowed only under extraordinary circumstances — substantially fewer than they currently are.

✓ **Responsiveness to changing social values should be a more important factor in federal land management**. It is difficult for managers to assess changing social values and how those values should affect their operations. Rule-based systems can provide guidance to more effectively integrate social values in natural resource decision making.

✓ **Federal agencies rarely have the capacity to promote and maintain community stability**. Many economic and social trends are beyond the control or capacity of federal resource management agencies to address. A more realistic policy would be for agencies to avoid programs and activities that disrupt local communities. In the event of necessary actions that could disrupt a local community, they should be phased in to mitigate social impacts.

✓ **Performance criteria for ecosystem management need to be established and used**. Because ecosystem management is relatively new and can significantly change resource management practices, it is important to assess the effectiveness of these approaches. In general, performance criteria for ecosystem management projects are lacking, and successes and failures are difficult to evaluate. Criteria should measure ecological, economic, and social performance. Expanding knowledge and experience about ecosystem approaches mean that performance criteria will need to be updated once established.

✓ **The past has much to tell us, if only we will heed it**. There is much to learn from our past, as well as that of other cultures that is useful in shaping viable public policies. Much writing of human history has tended to ignore the critical role that nature has played in shaping history and human values. In contrast, some of the natural sciences have largely ignored the role of people in shaping natural systems. Today's debate over natural resources is often driven by the polar extremes — by the so-called "anthropocentric" fringe, on the one hand, versus the "biocentric" fringe on the other. While these two extremes may seem worlds apart, they are both based upon a common theme and foundation that is fundamentally flawed: that nature and humans are separate and separable. Such excessively narrow views create a polarization, based on ideology that is very difficult to bridge in the arena of federal land management. The emerging fields of environmental history, human and cultural ecology and others can help bridge this gap. In practical terms, this means an explicit effort to describe the ecological and land use history of federal lands. If ecosystem management is to be successful, it must seek to "tell the story" of how the land and people came to be what they are today.

PRACTICAL CONSIDERATIONS

As with goals, the ways in which ecosystem approaches are implemented on the ground have significant differences and some overlaps with multiple-use management (see Table 2). Most notably, ecosystem management requires greater public input to integrate commodity and non-commodity uses across a larger area. Results are measured through indicators of long-term ecosystem health, such as species diversity and abundance, rather than output measures

Table 2. Comparison of Processes: Ecosystem Management and Multiple Use-Sustained Yield Management

Ecosystem Management	Multiple Use Management
1. Biological	*1. Biological*
Systems, integrated, or holistic approach, Multiple-use management or site by site consideration of both commodity and non-commodity resources	Multiple-use management of site-by-site consideration of both commodity and non-commodity resources
Nothing comparable	Management strategies organized in terms of individual resources
Landscape perspective	Site-specific perspective
Long-term temporal perspective	Sustained-yield management
Multiple scales of management	Stand-level management focus
Management within the range of natural variation	Management toward efficient production in terms of management objectives and the capabilities of the individual site
Species presence or populations as indicators	Resource outputs and inventories used as indicators
2. Socio-economic	*2. Socio-economic*
Ample use of public and private partnerships to accomplish programmatic objectives	Sparse use of partnerships to achieve specific programmatic objectives, e.g., fire suppression, insect and disease control, etc.
Systematic public involvement	Public involvement necessary to resolve site specific issues
3. Management	*3. Management*
Systematic interagency coordination	Interagency coordination sufficient to resolve site specific issues
Multidisciplinary management teams	Line-staff coordination among functional staffs encouraged

Fig. 1. The total U.S. production of wood, especially for lumber and fuelwood, climbed steeply from 1800 to 1900. Wood production stabilized between 1910 and 1970 as fuelwood was replaced by fossil fuels. After 1970, wood production started increasing again due to growing demand for paper and plywood. (Source: Fredrick &Sedjo, RFF, 1991.)

Domestic Production of Forest Products 1800–1985

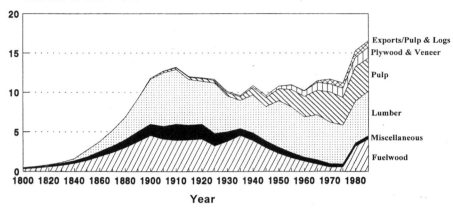

alone such as timber volume, visitor days, animal unit months (AUM), deer harvested, or trout caught.

Trends and patterns of human production and use of natural resources have undergone dramatic changes due to industrialization, population growth and increased per-capita consumption. These changes should be considered in developing ecosystem management strategies. In the past, some societies used resources in ways and quantities that were not sustainable. Archeological records from North America indicate that the Anazasi of Mesa Verde and other regions had seriously depleted their forest resources before the sites were abandoned. Within this century, wood was both the primary energy source and material basis for most people's lives (Fig. 1). After 1880, human needs for energy and materials have been increasingly supplied by underground resources as fossil fuels, minerals and metals have substituted for wood (Fig. 2). In New England,

Fig. 2. Wood has always been a vital resource in the United States, especially before the development of fossil fuels, steel and concrete.

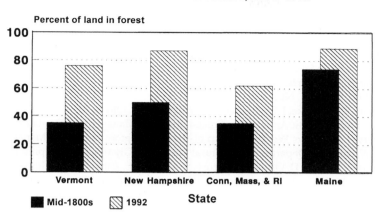

The Eastern Forest Comes Back
Trends in Eastern Forestland, 1850–1992

Percent of land in forest

■ Mid-1800s   ▨ 1992   State

Fig. 3. Forest lands in New England increased between 1850 and 1992 as agricultural lands were abandoned, and as industrial and residential users switched from wood to other sources of energy. (Sources: Harper, R.M. Changes in the Forestland of N.E. in Three Centuries. J. Forestry, 16: 442–52, 1918 and USDA/FS 1993a.)

forest areas dramatically depleted by agriculture, industry, and fuelwood use have rebounded dramatically during the past century (Fig. 3). However, concurrent increases in consumption and population have served to maintain human demands from and impacts on forest ecosystems.

There is no escaping the fact that translating concepts of ecosystem management into daily practice is a complex challenge. Managers must integrate across disciplines, political and ownership boundaries, and the diverging interests of a growing population. Resource managers — even if their responsibilities are within relatively small geographic boundaries — have to know the ecology of surrounding lands and waters, the social values of their neighbors and visitors, and the economic trends that drive resource uses in the area. For any individual, this is a herculean task. Reaching out to specialists and the public to form partnerships has become a fundamental part of the job and a defining characteristic of most ecosystem management projects.

A fundamental test of ecosystem management is whether it can help forge a working consensus on federal land management. To do that, ecosystem management must:

• Recognize the link between federal lands and regional and national economic and social systems;

• Define a more explicit role for resource production;

• Value historical knowledge and varied human roles in ecosystems;

• Recognize that judicious intensification of resource management for commodity production is both inevitable and desirable on some lands;

• Develop objective measures of ecosystem health and sustainability;

• Encourage further research especially related to ecosystem structure and function;

• Address the role of private lands in producing natural resources and maintaining ecosystem health;

• Improve federal agency abilities to collaboratively manage across agency boundaries and responsibilities, and;

• Be responsive to the varied and changing demands of the public for both commodity and amenity values.

SELECTED READING

Adams, D.A. 1993. *Renewable Resource Policy: The Legal-Institutional Foundations.* Island Press, Washington, DC.

Barbour, R.J., J.F. McNeel, S. Tesch, and D.B. Ryland. 1995. Management of Mixed Species, Small-diameter, Densely Stocked Stands. In: *Sustainability, Forest Health, and Meeting the Nations Needs for Wood Products.* COFE 1995 Council on Forest Engineering Annual Meeting, June 5–8, 1995, Cashiers, NC.

Cary, A.B., C. Elliott, B.R. Lippke, J. Sessions, C.J. Chambers, C.D. Oliver, J.F. Franklin, and M.J. Raphael. A Programmatic Approach to Small-Landscape Management: Final Report of the Biodiversity Pathways Working Group of the Washington Landscape Management Project. Washington Department of Natural Resources, Olympia, WA (in press).

Cissel, J., F. Swanson, G. Grant, D. Olson, S. Gregory, S. Garman, L. Ashkenas, M. Hunter, J. Kertis, J. Mayo, M. McSwain, K. Swindle, and D. Wallin. A Disturbance-Based Landscape Design in a Managed Forest Ecosystem: The Augusta Creek Study. U.S. Forest Service Pacific Northwest Station, Portland, OR (in press).

Cronon, W. 1985. *Changes in the Land: Indians, Colonists, and the Ecology of New England.* Hill and Wang, New York, NY.

Foster, D.R. 1993. *Land-use History and Forest Transformations in Central New England.* Springer-Verlag, Inc., New York, NY.

Frederick, K.D. and R.A. Sedjo (eds.). 1991. *America's Renewable Resources: Historic Trends and Current Challenges.* Resources For the Future, Washington, DC.

Interagency Ecosystem Management Task Force. 1995. *The Ecosystem Approach: Healthy Ecosystems and Sustainable Economies.* Vol. 1. National Technical Information Service, Springfield, VA.

Kohm, K.A. and J.F. Franklin, 1997. *Creating a Forestry for the 21st Century: The Science of Ecosystem Management.* Island Press.

Merlo, M. and Paveri. 1997. Formation and implementation of forest policies: a focus on the policy tools mix. *Policies, Institutions and Means for Sustainable Forestry Development.* Vol. 5. Proceedings of the XI World Forestry Congress, 13–22 October, 1997, Antalya, Turkey.

National Research Council. 1994. *Rangeland Health: New Methods to Classify, Inventory, and Monitor Rangelands.* National Academy Press, Washington, DC.

Salwerowicz, F. 1994. Mineral development and ecosystem management. In: Proceedings, Third International Conference on Environmental Issues and Waste Management in Energy and Mineral Production, Aug. 30–Sept. 1, Perth, West Australia. Curtin University of Technology.

Yaffee, S.L. 1994. *The Wisdom of the Spotted Owl: Policy Lessons for a New Century.* Island Press, Washington, DC.

Ecosystem Sustainability and Condition

Why Ecosystem Sustainability and Condition Are Important To Ecological Stewardship

For over a century, expanding concepts of sustainability — starting with principles of sustained-yield timber management — have driven changes in natural resource management policy and practice in the United States. Today, ecosystem approaches seek to integrate broad concerns about ecological, social, and economic sustainability into natural resource management. Human choices about what to sustain, where, why, and how have taken center stage in modern natural resource management. One of the toughest questions faced by natural resources managers is: What are the key elements of ecosystem sustainability and how are they affected by management practices? Sustainability is about the interplay of human values, goals, and behavior with biophysical ecosystem properties. Ecosystem structure and function, however, are key variables in sustainability because ecosystem conditions frame the choices humans make about what to sustain and why.

This summary was drafted by Robert Szaro with contributions from Ron Carroll, Ron Hendricks, and Myron Blank. It is based on the following chapter in Ecological Stewardship: A Common Reference For Ecosystem Management Vol. II:

Carroll, R., J. Belnap, R. Breckenridge, G. Meffe. "Ecosystem Sustainability and Condition."

KEYWORDS: ecosystem resilience, indicators, public participation, ecosystem disturbance

The fundamental requirement for sustainable ecosystem management is that the effects from human activities must not exceed the resiliency of the system to maintain essential functions and retain characteristic biodiversity. These constraints leave considerable management flexibility. Management for areas designated as "Wilderness Natural Areas", for example, will be strongly weighted towards maintaining natural ecosystem patterns and processes with minimal human influence. Other areas scheduled for timber harvest will, of course, include more economic considerations. In the latter case, ecosystem sustainability might only be achieved at a larger landscape scale in which timber harvests function ecologically as disturbance patches. Similar indicators of ecosystem condition might be used in both cases, but how they are interpreted will be influenced by the management goals for the area. The inherent dynamics of change in ecosystems makes rigid forms of management ineffective. Consequently, management must take a different tack: to be flexible, adaptive, and, as much as possible, predictable.

KEY FINDINGS

✓ **Historically, natural resource managers typically viewed nature as striving for equilibrium**. Ecosystems were long believed to have set stable points to which they returned after a disturbance. Single components of a system could be manipulated with predictable outcomes for the larger system. Early approaches to maximum sustained yield and single-species game management had such a perspective. Our contemporary perspective of ecosystems as dynamic, changing systems tells us that such prediction is difficult at best, probably impossible. Unfortunately, many resource management agencies proceed as though proper management can restore the "balance of nature" to its pre-disturbance condition after extraction of resources or other manipulations of the system. The target is seen as fixed, and the processes leading to that target are deemed to have predictable outcomes. Ecosystems are far from being this simple and predictable; management efforts must account for uncertainty, change, and surprise.

Fig. 1. Channeling of rivers and building levees to control "normal" floods can greatly exacerbate damage during catastrophic flood years (Photo courtesy of NOAA.).

✓ **Ecosystems have characteristic ranges of variability and resiliency to disturbances**. Managers must work within these ranges. Key biophysical properties of ecosystems must be addressed in management plans as well as in the general decision-making processes. Normal ranges of variation should not be exceeded, nor should the ecosystem be unduly constrained or variance too limited. Unduly constraining variation may lead to catastrophic changes: suppressing natural fires leads to destructive wildfires, controlling normal flood stages leads to catastrophic floods in unusually wet years (Fig. 1). Ideally, a managed ecosystem should include enough alternative habitats, seasonal food sources, successional sequences, and various refugia such that critical resources remain available somewhere during extreme years. The normal cycles and pulses of ecosystem dynamics must be maintained. Any particular ecosystem is linked to other ecosystems in a landscape context. Finding the right mix of spatial and temporal scales for establishing management boundaries is essential.

✓ **Ecosystems can recover from many kinds of disturbance but resiliency is limited**. Cumulative effects, persistent stresses, and catastrophic events must be avoided. Resiliency for any particular ecosystem should be investigated through field experimentation at the appropriate scale for the dominant management practices within the ecosystem.

✓ **Failure to distinguish between "ecosystem" as a research topic and "ecosystem" used to define a large-scale management unit often leads to confusion**. Researchers and managers often use the term "ecosystem" in very different ways. Researchers typically define ecosystem as a research construct at specific spatial and temporal scales to meet particular research objectives. Managers often use a more inclusive and geographically expansive definition to determine the boundaries and characteristics of a holistic management unit. Researchers and managers should be clear how they are using the term when communicating with each other.

✓ **People are integral parts of most ecosystems and the dominant players in some**. People — both consumptive and non-consumptive users of natural resources — must be treated both as agents that influence ecosystems and as stakeholders in any decision-making process that may affect the condition of the ecosystem (Fig. 2).

✓ **A working definition of ecosystem management with sustainability as its fundamental goal is helpful**. Carroll and Meffe (1995) developed the following definition: "Ecosystem management is an approach to maintaining or restoring the composition, structure, and function of natural and modified ecosystems for the goal of long-term sustainability. It is based on a collaboratively developed vision of desired future conditions and integrates ecological, socio-economic, and institutional perspectives, applied within a geographic framework defined primarily by natural ecological boundaries". This definition may seem cumbersome, but the concept is necessarily inclusive. Sustainable ecosystem management cannot succeed if ecological perspectives displace human welfare, if good ecological stewardship is sacrificed to meet narrow economic objectives, or if institutions cannot develop the flexibility to adopt better alternative perspectives.

✓ **Indicators of ecosystem condition must be reliable surrogates for both its ecological and social/economic attributes**. These indicators must be evaluated to determine their usefulness and predictability for measuring success and avoiding failure.

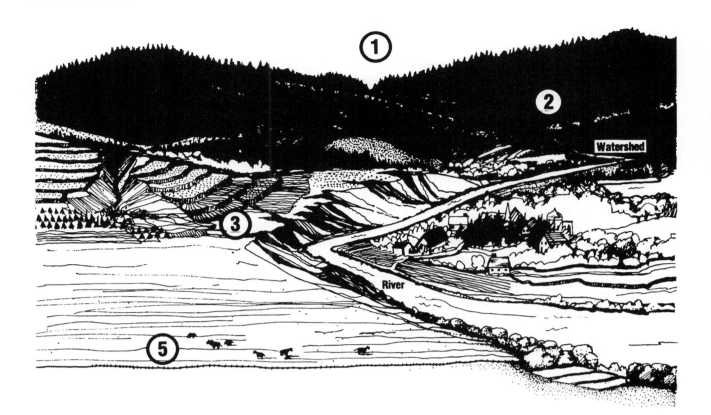

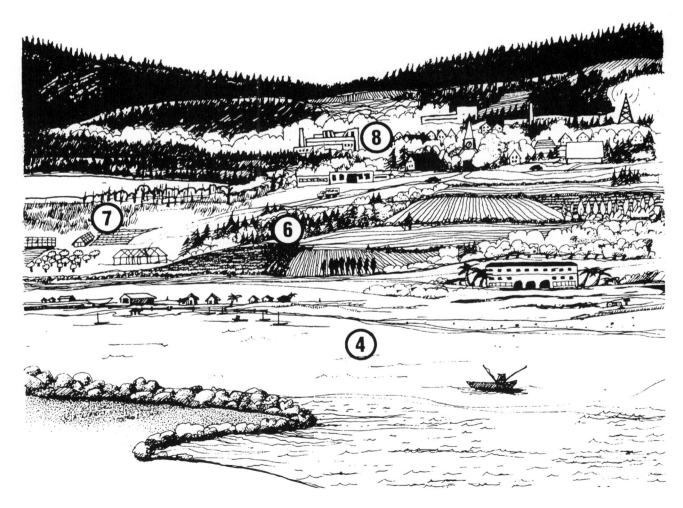

✓ **The goal of ecosystem management is to find approaches that meet human needs without degrading the environment**. In other words, sustainability is a primary goal of ecosystem management. Sustainability is not restricted to biophysical considerations but must also include the interplay of human values, goals, and behavior.

PRACTICAL CONSIDERATIONS

The social context for ecosystem sustainability. Efforts to manage ecosystems for their long-term sustainability take place in a social as well as an environmental context, with the costs and benefits unequally distributed in space and time. Because ecosystem management typically takes place on large landscape scales, stakeholder groups may be affected by management decisions in very different ways. For example, decisions to provide increased protection to biodiversity are likely to distribute social benefits broadly while concentrating costs locally. A more global distribution of people will value biodiversity for any reason, while local people bear most of the costs, usually in the form of opportunity costs when open access to hunting, fishing, extractable resources, or recreation are restricted. This spatial disjunction between costs and benefits makes consensus among stakeholders more difficult to achieve. Because a long time scale is implicit in sustainability, stakeholders may be asked to forego some immediate benefits to achieve long-term sustainability. Because people tend to discount the value of benefits to be realized in the future, it is difficult to explain the rationality of forgoing an immediate benefit, such as the pleasure some find in hunting keystone predators, in order to achieve the more diffuse benefits of a sustainable ecosystem for the future.

The culture of the agency bureaucracy. Successful ecosystem management for long-term sustainability requires agencies that are innovative, participatory, science-based, adaptive, and capable of long-term commitment of resources. Unfortunately, public land use agencies are bureaucracies with all their attendant decision-making inertia. To develop managers who can successfully pursue the process of ecosystem management, performance review criteria at all levels must be framed in ways that encourage managers to become risk-takers.

General properties of ecosystems. Particular ecosystems have unique properties critical to management decisions, but general properties shared by all ecosystems are also important to managers. Five such general properties that can, at least in principle, be characterized for any ecosystem are briefly described in Box 1.

Opposite: Fig. 2. Ecosystem approaches recognize that humans and their activities are an integral part of the ecosystem. In this idealized landscape, humans: (1) management wildlands for wildlife and biodiversity, and for carefully controlled harvest of timber and non-timber forest products; (2) manage watersheds from ridgetop to estuary for water quality and flow; (3) restore productivity to degraded lands; (4) manage coastal and marine areas for fisheries production and tourism; (5) manage grasslands for livestock and native biota; (6) practice diversified agriculture with integrated pest management and wildlife corridors; (7) develop community-based institutions to support sustainable natural resources management, andf (8) link residents of larger towns and cities with their environment through education, markets, and dialogues with natural resource managers and rural residents. (Courtesy of the World Resources Institute.)

Box 1
Management Significance of General Ecosystem Properties

Ecosystems are open systems: The functioning of any particular ecosystem is, to varying degrees, dependent on the surrounding environment for exchange of materials such as nutrients and waste products, energy input and release of dissipative energy, and for the colonization of species. Therefore the management of a particular ecosystem must recognize the various dependencies on other ecosystems and the surrounding environment.

Ecosystems are comprised of few linear and many non-linear processes: Populations grow geometrically and species interact with other species and with the physical environment in non-additive ways. Unexpected results and threshold responses will occur. Therefore, management decisions that could result in large and rapid changes to ecosystem functions should be avoided. Monitoring is essential.

Ecosystems have considerable internal spatial and temporal variations: The normal ranges of variation should not be exceeded nor should the ecosystem be unduly constrained and variance too limited. Unduly constraining variation will lead to catastrophic changes.

Ecosystems have characteristic trophic structures: Relatively simple food webs characterize extremely arid deserts, whereas complex webs are found in rainforests and coral reefs. In some ecosystems, the species composition in the higher trophic levels of food webs strongly influences net and secondary productivity and biodiversity at lower trophic levels.

Ecosystems are resilient: The resiliency properties of an ecosystem are ecosystem- and stress-specific; resiliency is not a generic property. Ecosystems that are degraded lose resiliency, especially where redundancy is reduced and nutrient transformation rates are slowed. Ecosystems can recover from many kinds of perturbations but resiliency is limited and cumulative effects, persistent stresses, and catastrophic events must be avoided.

Surprises, unexpected results, and threshold responses will occur. Monitoring the environmental responses to management decisions and maintaining the flexibility to modify decisions through adaptive management approaches are essential.

Management of large-scale ecosystems will require merging different approaches. First, a combination of information and analytical tools (historical studies, simulation models, geographic information systems,) should be used to develop predictive and alternative management approaches. Second, the decision-making process should be opened up by involving representative stakeholder groups to evaluate management alternatives.

Indicators of Ecosystem Condition — Measuring Success and Avoiding Failure. An ecosystem manager must have some indicators to measure the success or failure of decisions. These indicators must be reliable surrogates for both ecological and social/economic attributes, and they must be sensitive to degenerative or regenerative changes. The first step in identifying appropriate indicators is to classify the ecological resource as it currently exists (e.g., pinyon-juniper rangeland) and define the management boundaries. Once the ecosystem has been classified, its sustainable functions related to physical, biological, and social (aesthetics and economics) aspects can be identified based upon environmental and social values. Assessment questions can then be developed that focus on the specific ecosystem. For most ecosystems, sustainability and overall condition can be analyzed by subsuming assessment questions under the following broad categorical questions: (1) What is the biological integrity of the system? (2) How well does the system capture, store, and use water and energy? And (3) How efficient is the system at cycling nutrients? Indicators can then be selected that provide the connection between

---

Box 2
Evaluating the Condition of Public Rangelands in the Western United States

Stakeholders involved in evaluating whether rangelands are sustainable should first identify what values the ecosystems need to exhibit. For example, using the biological integrity of the ecosystem as a value would result in the development of assessment questions related to species composition and associated ecosystem functions. These would then be evaluated using a set of ecological indicators. The factors contributing to biological integrity can generally be extracted from previous research and then refined to meet local needs. Once the assessment questions have been developed, indicators can then be selected that answer or address the questions. For example, the assessment question, "Is the composition of vegetation at a particular site sustainable under current use?" could be addressed by selecting indicators such as plant species composition, ratios of native to exotic species, and amount of bare soil. These indicators would be most useful if they were made relative to reference ungrazed control sites that were otherwise similar. Other biophysical indicators might include measures of soil fertility, such as cation exchange capacity or amount of soil carbon, if loss of primary productivity was of concern.

---

ecological processes and the assessment questions. This framework makes it feasible to link values of various stakeholders with associated scientific issues and ecosystem function and structure (Box 2).

What makes a good indicator? Indicators are used to assess the condition of ecosystems because it is not feasible or cost-effective to measure every primary ecological variable. Selecting indicators useful for measuring change to major structural and functional aspects of the ecosystem provides a cost-effective and logical approach. There is a general set of characteristics for selecting indicators that have generic utility for most ecosystems, but will, of course, need to be refined to meet the site-specific needs of particular ecosystems (Box 3). Indicators can be grouped into three general areas related to physical, biological, and social impacts. These indices are often more useful in presenting information to the public. An index combines data from several measurements into information that is often used in trend analysis. An index can be derived for analysis of a change to a specific indicator such as vegetation, or can be used to assess change to a habitat type such as the condition of a riparian zone. If presented correctly and related in clear terms with examples, indices can provide an effective way to communicate concerns on complex issues. For example, a soil productivity index that combines information from indicators such as erosion, organic matter content, soil crust status, infiltration rate, and soil profile can be more easily understood by the general public.

SELECTED READING

Breckenridge, R.P., W.G. Kepner, and D.A. Mouat. 1995. A process for selecting indicators for monitoring conditions of rangeland health. *Environmental Monitoring and Assessment* 36: 45–60.

Carpenter, S.R., T.M. Frost, J.F. Kitchell, T.K. Kratz. 1993. Species dynamics and global environmental change: a perspective from ecosystem experiments. In: P.M. Kareiva, J.G. Kingsolver, R.B. Huey (eds.), *Biotic Interactions and Global Change.* Sinauer Associates Inc., Sunderland, Massachusetts

Box 3
Minimal Set of Characteristics for Evaluating Indicators of Ecosystem Condition

Characteristic 1

Ecosystem conceptual approach — the indicator should fit into an ecosystem approach that focuses on interactions of the system and not on a single isolated feature of the environment. To satisfy this characteristic, indicator parameters must relate in a known way to a structure or function of the ecological system to be evaluated so that the information obtained provides a "piece" of the overall puzzle.

Characteristic 2

Usability — the relative completeness and thoroughness of the procedure for measuring the indicator parameter provides the best gauge of indicator usability.

Characteristic 3

Cost-effectiveness — this characteristic can be evaluated by asking the question, "Is the incremental cost associated with the measurement low relative to the information obtained?"

Characteristic 4

Cause and effect — this characteristic focuses on whether there is a clear understanding of the mechanism of the relationship between changes in the degree of stress on the ecosystem and change in the value of the indicator. This characteristic can be evaluated by considering the following questions: (1) Does the indicator respond in a known, quantifiable and unambiguous manner to the stress in question?, (2) Is there dose–response information available for the indicator and the stress?, (3) Are exposure thresholds or trends known for the indicator?, and (4) Will the indicator provide similar information for most potential sampling areas within the same ecological unit?

Characteristic 5

Signal to noise ratio — This refers more specifically to the relative ease with which changes in the indicator may be distinguished from changes due to natural variability. To apply this criterion, the following questions should be considered: (1) Is the natural spatial and temporal variability associated with the parameter to be measured understood? (2) Are there predictable patterns in the spatial aspect (e.g., slope, soil associations, moisture) or temporal variability (e.g., seasonal) of the indicators identified?, and (3) Does the indicator possess significantly high signal strength in comparison to natural variability to allow detection of statistically significant changes within a reasonable time frame?

Characteristic 6

Alternate approaches — This characteristic would be satisfied if it can be concluded that no other approaches for measuring stress are available that would increase the quantity or quality of information obtained.

Characteristic 7

Quality assurance — This characteristic is satisfied if the quality of the resulting data can be reasonably assessed from a statistical and procedural standpoint. The process must be repeatable between different personnel and only requires a short training time.

Characteristic 8

Anticipatory — Ideally, an indicator for evaluating ecosystem conditions should be selected to provide an early warning signal of widespread changes in ecological condition or processes.

Characteristic 9

Historical records — Ideally, some historical data can be obtained for the parameter of interest from archived databases. Such data can be extremely valuable for establishing natural baseline conditions and the degree of natural variability associated with the parameter for the ecosystem. This information provides an important linkage between work that has been previously conducted and allows for trend assessment.

Characteristic 10

Retrospective — Some parameters allow for retrospective analysis in that new data may be generated that provide information on past conditions. For example, tree rings provide growth indices for each year of life of the tree. Because ecosystems are constantly under stress, it is important to try to evaluate how the stress has changed over time.

Characteristic 11

New information — Can the indicator being selected provide new information rather than simply replicating data that already exist? This is critical for identifying how new data can help meet the goal of better managing ecological resources.

Characteristic 12

Minimal environmental impact — Scientists and managers need to address the question of how much impact will the evaluation of an indicator impart on an environment. In fragile areas, such as desert systems, collecting an extensive amount of data or field sampling can often destroy the resource being protected. In a like manner, cutting down, harvesting, or removing parts of an ecosystem have to be considered in the overall evaluation of indicators for measuring change. Guidance in evaluating this characteristic should consider the question, "Does destructive sampling provide additional useful information that cannot be obtained without removal of the organisms being sampled?"

Carroll, C.R. 1996. Coarse woody debris in forest ecosystems: an overview of biodiversity issues and concepts. In: *Biodiversity and Coarse Woody Debris in Southern Forests*, Proceedings of the Workshop on Coarse Woody Debris in Southern Forests: Effects on Biodiversity; 1993 October 18–20; Athens, GA.

Franklin, J.F. 1995. Sustainability of managed temperate forest ecosystems. In: M. Munasinghe and W. Shearer (eds.), *Defining and Measuring Sustainability*. The United Nations University and The World Bank, Washington, DC.

Hann, W., M.E. Jensen, P.S. Bourgeron, and M. Prather. 1994. Land management assessment using hierarchical principles of landscape ecology. In: M.E. Jensen and P. S. Bourgeron (eds.), *Ecosystem Management: Principles and Application*. Vol. 2. U.S.D.A. Forest Service, Pacific Northwest Research Station.

Harwell, M.A., J.F. Long, A.M. Bartuska, J.H. Gentile, C.C. Harwell, V. Myers, J.C. Ogden. Ecosystem management to achieve ecological sustainability: the case of South Florida. *Environment Management* 20: 498–521.

Holling, C.S. 1995. Sustainability: The Cross-scale Dimension. In: M. Munasinghe and W. Shearer (eds.), *Defining and Measuring Sustainability*. The United Nations University and The World Bank, Washington DC.

Holling, C.S. and G.K. Meffe. 1996. Command and Control and the Pathology of Natural Resource Management. *Conservation Biology* 10: 328–337.

Hunsaker, C.T., D.B. Carpenter, J.J. Messer. 1990. Ecological indicators for regional monitoring. *Ecological Society of America Bulletin* 71: 165–172.

Noss, R.F. 1990. Indicators for monitoring biodiversity: a hierarchical approach. *Conservation Biology* 4: 355–364.

Odum, W.E., E.P. Odum, and H.T. Odum. 1995. Nature's pulsing paradigm. *Estuaries* 18: 547–555.

Paine, R.T. 1980. Food Webs: Linkage, interaction strength and community infrastructure. *Journal of Animal Ecology* 49: 667–685.

West, N.E. 1993. Biodiversity of rangelands. *Journal of Range Management* 46: 2–13.

Ecological Restoration

Why Restoration Ecology Is Important To Ecological Stewardship

Declines in species diversity, interrupted natural disturbance regimes, and the loss of naturally functioning ecological systems have become common resource management problems. A new specialization in ecological science, restoration ecology, has emerged to help reverse ecological degradation. Growing practical experience with ecological restoration in a range of ecosystems shows that management actions can help ecological systems recover. In fact, ecological restoration is at the center of some of America's most prominent ecosystem management efforts: the restoration of the Everglades, the expansion of late-successional forest habitats in the Pacific Northwest, and the rehabilitation of riparian habitats to restore endangered native fish populations in the desert Southwest (Box 1).

This summary was drafted by Nels Johnson and Jim Kenna with contributions from Wallace Covington, David Bayles, and Andrew Malk. It is based on the following chapters in Ecological Stewardship: A Common Reference for Ecosystem Management, Vol. II:

Covington, W., W. Niering, E. Starkey, and J. Walker. "Ecological Restoration and Management: Scientific Principles and Concepts."

Kenna, J., G. Robinson Jr., W. Pell, M. Thompson, and J. McNeel. "Ecosystem Restoration: A Manager's Perspective."

KEYWORDS: restoration ecology, reference conditions, rehabilitation, adaptive management, ecological degradation

---

Box 1. Restoration, Rehabilitation, or Reclamation?

Although restoration ecologists recognize a continuum of restoration actions, they often make distinctions between ecological *restoration* (recovery of natural systems to a range of conditions similar to those before significant degradation began), *rehabilitation* (recovery of biological productivity in a degraded ecosystem but not full restoration of species diversity or ecosystem processes), and *reclamation* (reestablishment of selected elements of the structure and function of severely degraded ecosystems). From a manager's perspective, each of these approaches has a role to play in regional ecological stewardship efforts.

Ecological restoration appears to provide a substantial common ground for implementing ecosystem management. Groups as diverse as the Society of American Foresters, the U.S. Fish and Wildlife Service, the American Forest and Paper Association, the Ecological Society of America, The Nature Conservancy, the National Academy of Sciences, and the United Nations have identified restoration as central to ecosystem approaches to natural resources management.

KEY FINDINGS

Restoration ecologists often recognize different types of approaches to repairing ecological damage. Notwithstanding these distinctions, *ecological restoration involves management actions designed to accelerate recovery of degraded ecosystems by complementing or reinforcing natural processes.* Ecological restoration is based upon principles of conservation biology and evolutionary ecology. Modern principles and concepts of ecological restoration are not entirely new — Aldo Leopold is generally credited for laying the foundations for restoration ecology as a discipline through his work to restore tall grass prairie and to manage the ecological impacts of deer populations in the forests of Wisconsin.

✓ **Ecological restoration should be considered when an ecosystem shows persistent signs of degradation.** Some sample indicators of ecological decline for terrestrial, riparian, and aquatic systems are shown in Box 2.

Box 2. Symptoms of Ecological Decline

Terrestrial Systems
- Dramatic changes in fire frequency and intensity (higher fuel loads) occur.
- Changes in species composition and vegetational structure are evident.
- The number of species at risk of extirpation or extinction increase.
- Invasive plants and animals have become abundant reducing the diversity, productivity, resilience, and abundance of native species.
- Widespread habitat change and fragmentation have reduced the habitat area and food resources for many species, and has diminished the ability of species and biological communities to migrate.

Riparian Systems
- The ability to absorb and moderate major flood events is reduced.
- Stream side vegetation and soil conditions have been substantially altered from reference conditions.
- Stream channel morphology (water velocity, water table) has changed.

- Water temperatures, nutrient loads, sediment loads, bacterial counts have increased (or decreased) beyond the range expected under reference conditions.
- The rate of degradation/erosion of stream banks has increased sharply.
- Patterns of extirpation or decline in sensitive aquatic populations are evident.
- Formerly widespread and abundant aquatic species are restricted to refugia.

Wetland Systems
- The regional availability of habitat for wetland dependent species is reduced.
- Water quality has dropped below a predefined historic level.
- Algal blooms are far in excess of expected natural levels.
- Native plant and animal communities have been altered or displaced by exotic species.
- Hydrogeomorphic processes have been altered.
- Soil erosion and deposition processes have changed unnaturally.

✓ **Ecological restoration uses an interdisciplinary approach to emulate natural ecosystem structure and function**. An interdisciplinary approach is vital since restoration efforts typically need to design and implement a broad array of practices. For example, ecosystem restoration in the ponderosa pine forests of the Southwest might consist of removing most trees which post-date Euro-American settlement, removing heavy fuels from the base of the old-growth trees, burning under prescription, removing introduced invasive weeds, and sowing with native herbaceous seeds. In riparian systems, practices might include bank stabilization, placement of instream structures, altering road design or location, reducing big game utilization of streamside vegetation, burning under aspen, planting riparian species, and changing livestock grazing, forest practices, or recreation use patterns (Fig. 1).

Fig. 1. Exclusion of grazing over a ten-year period has resulted in impressive regeneration of riparian vegetation and stabilization of stream banks along this stretch of the Crooked River in Central Oregon. (Photos courtesy of James Kenna.)

Box 3
An Approach to Adaptive Ecological Restoration

(1) Clearly diagnose the symptoms and causes of ecosystem degradation. What are the underlying mechanisms causing the symptoms?

(2) Determine reference conditions. What was the range of conditions of the ecosystem before substantial degradation?

(3) What factors — human and natural — are most limiting to the restoration process?

(4) Set measurable restoration goals and objectives, and relate them to reference conditions and limiting factors. How will you know if you are making progress towards those goals and objectives?

(5) Estimate outcomes of alternative treatments relative to limiting factors and reference conditions.

(6) Describe the planned treatment and the expected response.

(7) Design and implement a research or monitoring system to track ecosystem response to restoration activities so that success can be evaluated.

(8) Ensure mechanisms are in place to regularly integrate new information and "lessons learned" into planning and management activities.

✓ **Ecosystem restoration should not be construed as a fixed set of procedures, nor as a simple recipe for land management**. Rather, it is a framework for developing beneficial human/wildland interactions compatible with the range of natural variability (i.e., the last several thousand years) of native ecological systems. Ecosystem restoration is an exercise in complex problem-solving with an emphasis on natural processes and functions.

✓ **Defining the ecosystem reference conditions is a fundamental step in developing restoration plans**. The concept of the "evolutionary environment" (over a time scale of the past several thousand years) helps define reference conditions. In defining reference conditions, researchers typically seek to understand the frequency and intensity of disturbance regimes such as fire or flooding, the relative abundance and distribution of species and habitat types, the distribution and extent of seral stages across the landscape, interactions with humans, and environmental conditions such as soil moisture, nutrient availability, and hydrology. Important clues to the "evolutionary environment" can be found in paleobotanical records such as pollen sediments in lakes, geological records of flood events, dendochronological (tree ring) data, faunal remains associated with Native American archeological sites, and surviving tracts of relatively undisturbed habitats. However, the removal of Native American influences, the suppression of fire, and the addition of air and water pollutants may limit the usefulness of these tracts as reference areas. Therefore, knowledge of Native American resource use and management practices, together with historical records and observations made at the onset of Euro-American settlement, are important factors to consider in defining reference conditions.

✓ **Returning an ecosystem to a trajectory compatible with its "evolutionary environment," however, may not be a feasible or suitable goal**. Although agencies and organizations may target conditions and processes that fall within a range of variability (natural, historic or "reference"), they may also aim simply to improve certain conditions or control invasive species. For example, the U.S. and Canadian National Park Services sometimes target "pre-Columbian settlement" (pre-European) conditions as their restoration objective, while the Bureau of Land Management may target reference conditions based on the level of ecological function, rather than a point in time. In either case, understanding reference conditions and trends is essential to establishing restoration goals.

✓ **Successful restoration efforts depend on clear goals and the use of adaptive management**. Failure to clearly define the goal and objectives for ecosystem restoration has been a common problem in many efforts. Restoration in nearly all ecosystems must take place with limited information and considerable uncertainty about natural and human factors. For example, details about disturbance regimes, nutrient cycles, and species interactions are often poorly understood where restoration is needed. Therefore, the use of adaptive management practices in ecological restoration projects is strongly recommended. A systematic approach is essential for adaptive learning about ecological change and restoration (see Box 3).

✓ **For ecological restoration to be successful, collaborative partnerships among practitioners, researchers, landowners, and interest groups are essential**. An understanding of ecological processes and conditions is not the only key to successful restoration efforts. Regional restoration priorities may be driven by controversies, including endangered species listing, concerns over catastrophic disturbance regimes, and losses of unique habitats. In addition, some treatments may be inherently controversial and others untested.

Box 4
Common Features of Successful Restoration Projects

(1) Projects are organized around clear and practical objectives. A solid understanding of desired site conditions and trends has been developed.

(2) Physical and biological factors are integrated with social and economic considerations during design and implementation.

(3) An interdisciplinary technical team and any constituents necessary to successful implementation must be involved in the design of the project.

(4) Many projects must address combinations of ecosystem conditions. To do so effectively often requires a variety of actions to modify species composition, land uses, and ecosystem structure and processes.

(5) Monitoring is an important part of the design and management of the restoration project. Evaluating progress should emphasize the degree, direction, and rate of change rather than the attainment of quantitative benchmarks.

(6) A realistic assessment of physical, fiscal, and other constraints must be incorporated. This should include estimates of ongoing maintenance activities needed to ensure the success of the project.

(7) Restoration projects always encounter unpredictable or uncontrollable variables. For example, it may not rain as expected after a prescribed burn. Therefore, flexibility in implementation is important.

(8) Regular monitoring once implementation begins (and continuing after activities have ended) is vital to determine trends and allow for future management adjustments.

✓ **Economic costs and benefits play an important role in restoration decisions.** In some cases, restoration efforts can yield obvious and immediate benefits — increasing the quantity and quality of forage, controlling destructive weeds and introduced species, increasing water yield, controlling soil loss, and reducing public conflict over resource management. Still, restoration may require large up-front investments that may not yield economic benefits for years or even decades. Important lessons learned from restoration efforts around the United States are summarized in Box 4.

PRACTICAL CONSIDERATIONS

Restoration projects can be grouped based on the actions taken. There are three general types of restoration actions:

1. The *structural approach* includes engineered solutions designed to modify the present physical attributes of a system, or their arrangement and distribution, toward a more desirable state. For example, this may involve placing logs and rocks in streams to restore aquatic habitat diversity.

2. The *land-use approach* modifies the distribution, timing, intensity, or duration of uses affecting the landscape. For example, this may involve restrictions on grazing in sensitive riparian habitats.

3. The *biological control approach* changes the species composition of an ecosystem, controlling undesirable species or introducing and promoting desirable ones. Controlling leafy spurge in the native grasslands of Theodore Roosevelt National Park or reintroducing wolves in the Greater Yellowstone Ecosystem are examples of this approach (Fig. 2).

Fig. 2. The reintroduction of a keystone predator, such as the wolf in Yellowstone National Park, can have major consequences for ecosystem composition. (Photo courtesy of U.S. Fish and Wildlife Service.)

Usually more than one of these approaches is used in a given restoration project, although one may be predominant. Frequently, the need to modify land uses is dominant.

Deciding when restoration is needed is an important issue. A sample of decision factors is presented in Box 5. Reclamation and rehabilitation may be more appropriate than restoration in areas where ecosystems have been severely degraded and few biological legacies from the original natural ecosystem remain. These areas may include abandoned mine lands, eroded agricultural lands, vacant or contaminated urban and industrial sites, and areas affected by severe natural disturbance, such as large floods, landslides, and volcanic eruptions (Fig. 3). In many cases, native vegetation, soil microbes, and animal communities have been killed or removed, and most of the topsoil may have been lost, altered, or buried. Even limited reclamation or rehabilitation on such sites can be extremely expensive.

Box 5
Restoration Decision Factors

All restoration projects require decisions about the commitment of time, effort and money. The following list provides some common factors that help decide what actions are most important to take first.

- *Existing Level of Ecological Function.* Preventive measures in systems that are functioning, but at risk, can be very good investments.

- *Cost Effectiveness.* Different actions will generally yield different degrees of restoration relative to their cost.

- *Conservation Values.* Various species and communities may be valued more highly due to their rarity, sensitivity, social role or economic importance.

- *Potential Effects.* Some actions may be more likely to yield the desired response due to limiting factors, land ownership patterns, public support, or potential barriers.

- *Relationship to Other Actions and Trends.* Land uses, management projects and resource trends will affect the success of restoration projects.

- *Risk Assessment.* Available experience and knowledge helps determine the likelihood of success and the risk of failure.

Fig. 3. Acid drainage from abandoned mines can impair water quality so severely that reclamation or rehabilitation is a more realistic goal than restoration. (Photo courtesy of James Kenna.)

Ultimately, the decision to proceed with a restoration project rests on its feasibility and its priority. Some of the issues to consider in evaluating the feasibility of a restoration project include the following:

1. Can any options be excluded because of low probability of success?

2. Is the cost of any option (including cost to society) prohibitive?

3. What time will be required to reach each of the goals?

4. Is further disturbance likely to occur that will limit the effectiveness of restoration efforts?

5. Will restoration efforts satisfy regulatory requirements?

6. Is cooperation with third parties or stakeholders possible?

7. Are resources available for adequate monitoring of restoration success?

SELECTED READING

Apfelbaum, S.L. and K.A. Chapman. 1994. Ecological restoration: a practical approach. In: T.R. Crow, A. Haney and D.M. Waller (eds.), *Ecosystem Management*. GTR NC-166. USDA Forest Service, North Central Forest Experiment Station, St. Paul, MN.

Berger, J.J. (ed.). 1990. *Environmental Restoration — Science and Strategies for Restoring the Earth*. Island Press, Covelo, CA.

Cairney, T. (ed.). 1987. *Reclaiming Contaminated Land*. Blackie and Son, Ltd., London.

Elmore, W. and B. Kaufman. 1994. Riparian and watershed systems: degradation and restoration. In: M. Vavra, W.A. Laylock and R.D. Piper (eds.), *Ecological Implications of Livestock Herbivory in the West*. Society for Range Management, Denver, CO.

Jaindl, R.G. and T.M. Quigley (eds.). 1996. *A Search for a Solution: the Land, People, and Economy of the Blue Mountains.* American Forests in cooperation with the Blue Mountains Natural Resource Institute, Washington DC.

National Research Council. 1992. *Restoration of Aquatic Ecosystems: Science Technology and Public Policy.* S. Maurizi and F. Poillon (eds.). National Academy Press, Washington DC.

Pritchard, D., H. Barrett, K. Gebhardt, J. Cagney, P.L. Hansen, R. Clark, B. Mitchell, J. Fogg, D. Tippy. 1993. *Riparian Area Management: Process for Assessing Proper Functioning Condition.* USDI Bureau of Land Management, Denver, CO.

Pritchard, D., C. Bridges, S. Leonard, R. Krapf and W. Hagenbuck. 1994. *Riparian Area Management: Process for Assessing Proper Functioning Condition of Riparian-Wetland Areas.* USDI Bureau of Land Management, Denver, CO.

Quigley, T.M. and S.J. Arbelbide. 1997. *An Assessment of Ecosystem Components in the Interior Columbia Basin and Portions of the Klamath and Great Basins.* Vol. 1. USDA Forest Service, Pacific Northwest Research Station, Portland, OR.

U.S. Congress, Office of Technology Assessment. 1993. *Harmful Non-indigenous Species in the Untied States.* U.S. Government Printing Office, Washington DC.

Public Expectations, Values, and Law

♦ *Legal Perspectives*

♦ *Public Expectations and Shifting Values*

♦ *Evolving Public Agency Beliefs and Behaviors*

♦ *Processes for Collaboration*

♦ *Regional Cooperation*

Legal Perspectives

Why Legal Perspectives Are Important To Ecological Stewardship

No federal law mandates "ecosystem management." In fact, most of the laws governing the use and management of public natural resources were passed decades ago, long before concepts of ecosystem management began to emerge. Although the existing law does not expressly provide for ecosystem management, it contains substantial legal authority to support new ecosystem management policies and initiatives. In combination, the organic legislation governing the land management agencies along with laws like the National Environmental Policy Act (NEPA) and the Endangered Species Act (ESA) provide a strong legal foundation for emerging federal ecosystem management efforts (Table 1). The combined impact of these laws may virtually require federal land managers to engage in ecosystem management to meet current legal obligations. Thus, there are legal as well as social and scientific justifications for moving toward an ecosystem approach in natural resources management. Other laws, including the Federal Advisory Committee Act (FACA), resource development statutes, and takings limitations, impose important constraints that may temper — but should not preclude — new federal ecosystem management policies.

KEYWORDS: Law, legislation, courts, litigation, administrative policy

This summary was drafted by Nels Johnson with contributions from Robert Keiter, William Snape, Ronald Kaufman, and Andrew Malk. It is based on the following chapter in Ecological Stewardship: A Common Reference For Ecosystem Management, Vol. III:

Keiter, Robert B., Ted Boling, and Louise Milkman. "Legal Perspectives on Ecosystem Management: Legitimizing a New Federal Land Management Policy."

Table 1. The Federal Legal Framework for Ecosystem Management

(1) **"Organic" Acts** (principal boundary-based laws governing the federal land management agencies):

• National Parks Organic Act

• The Wilderness Act (applies to multiple agencies)

• Multiple Use–Sustained Yield Act

• National Forest Management Act

• Federal Land Policy and Management Act

• National Wildlife Refuge System Administration Act

• Wild and Scenic Rivers Act (applies to multiple agencies)

(2) **Cross-Cutting Acts** (principal laws that apply across jurisdictional boundaries):

• National Environmental Policy Act — administered by action agency with oversight by White House Council on Environmental Quality

• Endangered Species Act — administered by the US Fish and Wildlife Service and National Marine Fisheries Service.

• Clean Water Act — administered by Environmental Protection Agency, Army Corps of Engineers, and the states.

• Federal Advisory Committee Act — administered by action agency with oversight by General Services Administration.

Understanding the legal context for ecosystem approaches in natural resources management is important for several reasons. First, some laws impose various substantive obligations and prohibitions on resource managers and decision makers. For example, the Clean Water Act and the ESA have specific prescriptive requirements that natural resource managers must observe while carrying out their routine responsibilities. Second, other laws, such as NEPA, the Federal Land Policy and Management Act (FLPMA), and FACA, establish procedural obligations that managers must follow when planning or evaluating management options. Some laws, such as the National Forest Management Act (NFMA) and the Federal Land Policy and Management Act (FLPMA) impose both substantive and procedural requirements. Third, the legal system is a flexible framework that changes over time as agencies develop and revise rules and regulations to implement laws and courts rule on conflicts between laws or reject the manner in which legal provisions are being implemented. Managers and decision makers should be aware of how these changes might affect their responsibilities. Fourth, agencies have differing authorizing legislation or organic acts which set out their responsibilities. This legislation (and subsequent amendments) can provide a "roadmap" that helps define roles and responsibilities for interagency collaboration across large landscapes, including between federal and state agencies. Finally, knowledge of legal requirements can provide an important catalyst for innovation in natural resources management. For example, Habitat Conservation Plans (HCPs) are an innovative, if controversial, strategy for protecting endangered species on private lands while minimizing legal restrictions on land owners.

KEY FINDINGS

To the uninitiated, the legal framework governing natural resources management in the United States is a complex and tangled web. Nevertheless, this legal framework enables more often than it hinders managers seeking to implement an ecosystem approach to natural resources management. A very basic understanding of the following points should help resource managers to understand the legal context in which they work.

✓ **Because of the integrated nature of ecosystem management, no one law is fully applicable to its implementation.** For nearly all federal natural resource managers, broad federal legislation such as NEPA and FACA, and often ESA, the Clean Water Act, and sometimes the Clean Air Act, will be relevant to planning and implementing ecosystem approaches. In addition, managers will want to be aware of how their agency's organic act and other specific laws — the National Forest Management Act (NFMA), the Rangeland and Forest Resources Planning Act (RPA) and the Wilderness Act in the case of the U.S. Forest Service — create both opportunities and legal boundaries for what they can and cannot do to pursue an ecosystem approach. Given that many ecosystem management efforts cross jurisdictional boundaries, laws specifically governing the mandates and functions of other agencies will likely be relevant as well.

✓ **The various laws that guide ecosystem management can sometimes be in conflict.** For example, Yellowstone National Park officials recently faced a court challenge over their decision to retain the Fishing Bridge campground, which is located in prime grizzly bear habitat, to meet visitor needs. In this case, the National Park Service was caught between its organic act, which emphasizes preserving natural features for the enjoyment of visitors, and the ESA, which emphasizes protecting endangered species and minimizing human threats to their recovery. In other instances, adjacent land management agencies operating under different legal mandates can find themselves in conflict over shared ecological resources. In and around Yellowstone National Park, for example, the National Park Service's preservationist goals have sometimes conflicted with the multiple-use mandates governing the U.S. Forest Service and the Bureau of Land Management that are responsible for adjacent lands. Under an ecosystem approach, resource managers must address conflicting mandates directly and jointly to ensure the ecological integrity of the shared resource base.

✓ **To implement ecosystem management initiatives, federal managers must often be acquainted with state, local, and tribal planning and zoning ordinances.** Federal familiarity with these processes should provide opportunities for cooperative ecosystem management efforts among federal, state and local authorities. Knowledge of the legal roles and responsibilities of state and local authorities can also help identify local agencies that should be involved in planning efforts and might be in a position to take on certain implementation or monitoring responsibilities.

✓ **Legal requirements can affect private and public land managers differently.** Private land managers, for example, are not bound in their planning by the procedural requirements of NEPA and FACA while federal land managers are. Other laws which impose certain obligations on federal managers may be implemented differently on private lands. For example, federal land managers can use incentive mechanisms, such as Habitat

Fig. 1. Compliance with Clean Water Act regulations to protect wetlands requires cooperation between federal and state agencies and private property owners. (Photo courtesy of James Kenna.)

Conservation Plans (HCP) under the ESA, when they work directly with private landowners. A district ranger on a National Forest, however, cannot use an HCP to avoid future restrictions on resource management activities in his or her district. Under the Clean Water Act, cooperation between federal agencies and private landowners is also necessary for wetlands protection and other permitting decisions (Fig. 1).

✓ **The scale of ecosystem management assessment and planning involves important legal issues**. Under an ecosystem approach, federal planning and management are performed at a variety of geographic and temporal scales. The principal levels are ecosystem-wide planning, plan-level management, and project or site-specific decisions. Adaptive management entails change and adjustment at each level to address new information.

Box 1. When and Where Do Legal Requirements Start?

The Ninth Circuit's 1994 decision in *Pacific Rivers Council v. Thomas*, 30 F.3d 1050 (9th Cir. 1994), illustrates some of the difficult NEPA and ESA compliance issues that ecosystem management initiatives can raise. In this case, the court viewed federal land management plans as agency actions with long-lasting effects that required reinitiated consultation upon the ESA listing of a species. Although the court did not rule on whether any action that "may affect" the species must be suspended pending further consultation (16 U.S.C. § 1536(d)), it noted that timber sales had previously been treated as *per se* irreversible and irretrievable commitments. Lower courts subsequently ruled that timber sales and other site-specific projects must be enjoined until plan-level consultation is complete. The decisions have significant legal implications for ecosystem-wide planning and adaptive management, namely that land management plans have environmental impacts that are independent of the project decisions that may occur in the course of plan implementation. Under NEPA, the questions are when to prepare EIS documents, at what depth, and when is supplementation required. Under the ESA, the questions are when to consult with the U.S. Fish and Wildlife Service and the National Marine Fisheries Service and when reconsultation is required. And, under the land management statutes, the question is how to integrate NEPA and ESA compliance with traditional planning processes, such as national forest management plans mandated under NFMA and typically intended to last for a decade or more. Responsible land managers must ensure that resource management decisions are in compliance with NEPA and ESA requirements at all planning levels.

Different legal requirements may apply at each level, depending on the scope and nature of the agency action. A major legal challenge confronting land management agencies pursuing ecosystem management programs is to harmonize and coordinate these legal obligations, especially those arising under NEPA and ESA (Box 1).

✓ **Given the myriad and sometimes even contradictory laws governing public lands and resources, legal complexity will surface as an ecosystem management concern.** Judicial challenges to ecosystem management can be expected and should be viewed as an opportunity to clarify agency authority and responsibility under present law. Because the courts are available to review ecosystem management policies, judicial decisions can either legitimize or overturn such policies (Box 2). However, courts traditionally give considerable deference to administrative expertise and discretion (see below under Practical Considerations). The dynamic nature of ecosystems and shifting social values necessitates a flexible legal framework to implement ecosystem management.

Draft Supplemental Environmental Impact Statement

on Management of Habitat for Late-Successional and Old-Growth Forest Related Species Within the Range of the Northern Spotted Owl

Fig. 2. Judicial decisions relating to the adequacy of Environmental Impact Statements, including this one on the northern spotted owl, have provided a legal context for ecosystem management.

Box 2
Going to Court: a Perspective on Litigation

Although federal agencies ordinarily seek to avoid litigation, some litigation over new ecosystem management policies is inevitable. Whenever a significant policy shift occurs, it will be subject to challenge in court, as already has occurred with the President's Forest Plan in the Pacific Northwest (*Seattle Audubon Society v. Lyons*, 871 F.Supp. 1291 (W.D. Wash. 1994), *affirmed*, 80 F.3d 1401 (9th Cir. 1996)). Litigation can sometimes be a counterproductive use of resources, and it can undermine the cooperative relationships that are critical to the success of an ecosystem management approach. However, the threat of litigation, should not deter land managers from pursuing bold ecosystem management initiatives within the rather broad scope of their existing legal authority. When litigation cannot be avoided, then it should be viewed as an opportunity to define the limits of agency authority and to clarify where legislative assistance may be required.

✓ **Within the existing legal framework, the appropriations process can be used to support and retard ecosystem management initiatives.** Without sufficient funding, federal agencies and others cannot fully employ their legal authority to implement ecosystem management initiatives. Moreover, substantive riders to appropriations bills that contain short-term amendments to existing environmental laws can significantly change the governing legal structure and adversely affect ecosystem management initiatives.

PRACTICAL CONSIDERATIONS

The law presents challenges to ecosystem management approaches. Many of these challenges, however, can be addressed within the existing legal framework.

When litigation challenges the use of ecosystem management approaches, agencies can clarify the legal basis for ecosystem management in several ways. For example, public land management agencies can use their existing rule-making authority to promulgate regulations that would further clarify ecosystem management policy. Executive orders and Solicitor's opinions might also be used for this purpose. Under the current legal framework, the goal should be to establish clear ecosystem management standards that would promote clarity and certainty among those who will be affected. This should also promote agency accountability and perhaps dissipate concern over ecosystem management among traditional public land constituencies. And, it would further legitimize ecosystem management on the public lands.

Although the Fifth Amendment takings clause has been raised as an obstacle to ecosystem management initiatives, recent Supreme Court takings decisions have only limited relevance to most federal ecosystem management initiatives. For the most part, the Court's decisions have involved cases where local governing bodies significantly limited an owner's development options on privately owned land. In contrast, because public lands and resources are not privately owned, property rights can arise only when the government grants them. In the ecosystem management context, the takings doctrine would have the most effect if public land managers sought to extend federal regulatory power to adjacent private lands to protect ecosystem resources.

The Federal Advisory Committee Act (FACA) has also been cited as an obstacle to ecosystem management. FACA challenges have hindered several federal ecosystem management efforts. The perception that FACA prevents or limits the participation of non-governmental officials has also restricted federal ecosystem management efforts. The FACA imposes procedural requirements on federal agencies in certain circumstances when the agencies solicit and receive collective advice from persons who are not full-time federal employees. In several instances, courts have ruled that federal agencies may meet with or accept non-consensus advice from non-federal parties. In addition, Congress amended FACA in 1995 to exempt meetings that are held between federal officials and officials from state, local, and tribal governments. Nonetheless, the FACA represents one specific area where policy clarification could further facilitate ecosystem management through early, constructive, and comprehensive public participation.

Certain long-standing laws and legal doctrines do pose difficult challenges to ecosystem approaches. For example, the General Mining Law of 1872 and the state-based prior appropriation doctrine governing water allocation in the West give resources users powerful property rights that are based entirely upon priority of use with minimal or no consideration given to countervailing ecological and social concerns. In these cases, economic strategies rather than legal strategies may be needed to accomplish ecosystem management objectives.

The current legal framework is sufficient for a shift toward ecosystem approaches, although certain changes in law may be desirable. Because existing public land and environmental laws were not drafted with ecosystem management in mind, it may be necessary to seek changes in the law to achieve some ecosystem management goals. Under our constitutional structure, Congress has the authority to make and change law. Whenever the law prevents the federal agencies from implementing a new ecosystem management policy, consideration should be given to enlisting Congress to change the law. Although pursuing new legislation entails some risk, carefully targeted legislative initiatives can reduce that risk. Congressional assistance already has been enlisted to amend FACA to exempt intergovernmental partnership arrangements from rigid procedural requirements. Before pursuing new legislation, however, a determination must be made whether the legislative proposal should be designed to respond to an individual agency's specific problem, or whether broader statutory changes applicable to all agencies should be sought. Still, the history of public land law reveals that new policy shifts, which have occurred regularly and which now include the concept of ecosystem management, ordinarily are implemented and tested administratively before changes are confirmed in law. That process has now begun and should be pursued under the existing legal regime. The law provides federal resource management agencies with sufficient legal authority to begin making the shift to ecosystem management on the public domain.

SELECTED READING

Bosselman, F., and D. Tarlock. 1994. Symposium on Ecology and the Law. *Chicago-Kent Law Review* 69: 843–985.

Center for Wildlife Law. 1996. *Saving Biodiversity: A Status Report on State Laws, Policies and Programs.* Defenders of Wildlife, Washington, DC.

Houck, O.A. 1997. On the law of biodiversity and ecosystem management. *Minnesota Law Review* 81: 869–979.

Interagency Ecosystem Management Task Force. 1995a. *The Ecosystem Approach: Healthy Ecosystems and Sustainable Economies.* Vol. I: Overview. Washington, DC.

Interagency Ecosystem Management Task Force. 1995b. *The Ecosystem Approach: Healthy Ecosystems and Sustainable Economies.* Vol. II: Implementation Issues. Washington, DC.

Keiter, R.B. 1994. Beyond the boundary line: constructing a law of ecosystem management. *University of Colorado Law Review* 65: 293–333.

Keystone Center. 1996. *Keystone National Policy Dialogue on Ecosystem Management: Implementing Community-Based Approaches.* Keystone Center, Keystone, CO.

U.S. General Accounting Office. 1994. *Ecosystem Management: Additional Actions Needed to Adequately Test a Promising Approach.* GAO/RCED-94-111. Washington, DC.

Public Expectations and Shifting Values

Why Public Expectations And Shifting Values Are Important To Ecological Stewardship

Public opinion has always been important in shaping natural resource management in the United States. Public sentiment helped create the national forest system a century ago and provided the groundswell behind the landmark environmental legislation of the 1970s. In an earlier era, however, public opinion was easier to account for. In general, people had more convergent views than exist today. A rough national consensus existed for decades about the need for a utilitarian approach to conservation, such as that espoused by Gifford Pinchot. John Muir and the preservationists provided a distinctive but minority view. In practice, special interests often held sway because the public had relatively little access to policy makers and it was difficult to assess public views on specific issues. Times have changed. Now, the public is much more engaged in what happens to their natural resources and the management of public lands and waters. Legal and administrative provisions have created more opportunities for the public to be involved in planning and decisionmaking. Still, public views are perhaps more divergent about resource management priorities than ever. Public agencies have often failed to keep pace with the rapid evolution of public views, or to seek ways to reconcile their divergence. As a result, resource management agencies have found themselves mired in lawsuits, legal and administrative appeals, and intense political controversies.

KEYWORDS: Public opinion, research methods, social science

This summary was drafted by Andrew Malk with contributions from John Bliss, Gerald Helton, and Bill Shaw. It is based on the following chapters in Ecological Stewardship: A Common Reference for Ecosystem Management, Vol. III:

Bliss, John. "Understanding People in the Landscape: Social Research Applications for Ecological Stewardship."

Cordell, Ken, Howie Thompson, Barbara MacDonald, Clyde Thompson, Cristina Ramos, Gerald Helton, Stephen Ragone, and Michelle Dawson. "Shifting Values and Public Expectations."

Box 1
What People Mean When They Say They Are Environmentalists: a Social Science Study

Recent surveys report that well over half of Americans consider themselves to be "environmentalists." But what does this mean? Anthropologists Willett Kempton, James Boster, and Jennifer Hartley utilized a combination of qualitative and quantitative research methods to discover the underlying meanings of American environmentalism. They began by conducting interviews with citizens from various walks of life. Through these interviews, they learned how laypeople think about environmental issues — the words they use, the issues that concern them, the relationships they see between the issues. To gain a deeper understanding of how these issues and values affect individuals' decisions, the researchers developed case studies of selected interviewees actively involved with environmental policy issues: two congressional staff members, an environmental activist, and a mining union representative.

They also administered a survey to five groups chosen to represent diverse viewpoints on environmental matters: a California Sierra Club chapter, participants in Earth First! meetings in Wisconsin and Vermont, managers of dry cleaning shops (a business heavily impacted by clean air legislation) in Los Angeles, laid-off sawmill workers in Oregon, and a random sample of the general public in California.

Through this innovative combination of methodologies the research team gained insights into how opinions about environmental issues are formed, how they relate to other beliefs and values, and how they are distributed among various publics. Study results suggest that "environmentalism has already become integrated with core American values such as parental responsibility, obligation to descendants, and traditional religious teachings." They concluded that "American environmentalism represents a consensus view, its major tenets are held by large majorities, and it is not opposed on its own terms by any alternative coherent belief system."

Social scientists have developed a range of methods to identify and describe public expectations and values. These methods can help public agencies to keep pace of public views and be more responsive to their expectations. They can be used to gather information across a range of interests and social groups and from national to local scales. They can also help researchers and natural resource managers get a more precise idea of what popular terms really mean. For example, Box 1 shows how a study using a variety of research methods was able to describe what people mean when they label themselves "environmentalists".

KEY FINDINGS

There are several key questions any researcher or manager should ask before selecting a method to evaluate public opinions and values on natural resource issues.

✓ **What is the research objective?** Clearly stated objectives will greatly facilitate an efficient, effective research effort; ambiguous objectives will surely doom it. Involving a social scientist during this first step will help frame the right question(s) and choose the right sampling technique. Research objectives may be stated as testable hypotheses or as exploratory questions. Using a deductive model, the researcher formulates hypotheses regarding the phenomenon of interest, tests the hypotheses with relevant data, and either accepts or rejects the hypotheses. If the objective is to gain a basic understanding of general characteristics of a large population — such as demographic information about park visitors — a deductive method is likely appropriate. Inductive research starts from the more humble position of admitting ignorance as to what hypotheses might be appropriate. Rather than testing a predetermined hypothesis, the researcher gathers and sifts through potentially relevant data, searching for explanatory patterns. Only then is a testable hypothesis selected.

✓ **About whom do we wish to learn?** The size of the population to be sampled determines the sampling method. National public opinion polling requires a very different set of tools, expertise, and investment than does convening a focus group of local opinion leaders. Therefore, coincident with clarifying the research objective, one must consider the question "About whom do I wish to make inferences?". If the study sample doesn't represent the population of interest, neither will the results. For example, assessing public expectations for Rocky Mountain National Park would likely include both sampling of local residents and a national sample of the general public, since people from all parts of the country visit the park. However, changing Bureau of Land Management schedules for closing roads before winter sets in on remote Steens Mountain in eastern Oregon area might only entail sampling several counties near the area. Assumptions about who is or is not represented in the sample need to be carefully examined. For example, participants in so-called "public" meetings are not often representative of many views that exist in a community (Box 2).

✓ **Who will do the research?** Social science research consists of more or less formalized methods for learning about people, their opinions, beliefs, and behavior. This learning can take place on a number of levels of rigor, complexity, and sophistication. The manager who takes the time to get out of the pickup and become acquainted with local people is conducting unstructured social research. More formal social science learning may be conducted by agency personnel with the assistance of professional researchers. Large-scale or long-term research efforts will benefit from direct professional involvement, as will complex or sensitive problems. In cases like these, a trained social scientist can bring greater methodological and analytical expertise as well as greater impartiality. A second source of expertise is academia. Most faculty members are continuously on the look-out for challenging research projects for themselves and their graduate students. Cooperative agreements between federal agencies and universities are common. Participatory research, in which concerned citizens, organizations, and researchers cooperate to educate themselves about questions of mutual concern, is another strategy. The defining characteristic of participatory research is that people most affected by the phenomena under study participate directly in problem definition, data collection and analysis, and utilization of study results to solve local problems.

> **Box 2**
> **How Public are Public Meetings?**
>
> It is sometimes incorrectly assumed that participants at a public meeting constitute a random sample of those interested in the subject of the meeting, and that their opinions are representative of those of the larger population. This is rarely the case, as public meetings attract only those persons who are aware of the meeting (i.e. read the newspaper or watch the news), able to attend (have time and transportation), and motivated to learn about or share their views on the issues addressed by the meeting. In most cases, it is better to think of such events as "open" meetings rather than "public" meetings where fully representative views are expressed. Such meetings, however, are often attended by the people who are the most active and interested in the issues. Their views may be disproportionately important, but they should not be assumed to be representative.

PRACTICAL CONSIDERATIONS

A wide array of social science research methods can be used to assess public views and values on the natural resource management. Each has advantages and limitations, depending on the purpose it is being used for. Table 1 briefly summarizes some of these tools, while Fig. 1 indicates how a researcher might

Table 1. Illustrative list of social science research methodologies.

Method	Application	Limitations
Secondary Data –Historical –Census –Previous research	Provide context and background for any social research endeavor	Availability and relevance of existing data
Ethnographic –Participant observation –Case study –Oral history –Key informant	Explain experience and values of specific target population, identify relationships, understand issues in context	Time requirement, limited capacity to generalize, lack of formal analytical procedures
Structured Group –Focus group –Nominal group –Delphi	Establish problem's boundaries and topics for further research	Limited capacity to generalize
Survey –Telephone –Mail –Door to door	Estimate general parameters of large population, rigorous statistical analysis	A priori knowledge required, limited capacity to explain, declining response rates

select an appropriate social science tool to assess public views. In some cases, existing information, or secondary data, may answer the relevant questions. When research is needed to collect new information, researchers must determine whether quantitative or qualitative methods, or some combination, are most appropriate.

Secondary Data. Any thorough research begins with a trip to the library. Before investing a great deal in a new research project, it's worth investing a little in discovering what others have already done. When looking for comparable historical studies, the closer the scale and subject matter is to the issue at hand, the better. A broadly useful type of secondary data is census data, which for the US cover demographics, employment, occupation, income, leisure, health, migration, and a wide variety of other population and individual traits. Before embarking on an extensive search for data, project staff should spend considerable time thinking about why the data are needed and how they will be used.

Quantitative Survey Research. A properly designed survey is an effective, relatively rapid means to quantify population characteristics of interest. A survey approach should be considered if: (1) the questions of interest, and the range of appropriate responses, are already known to the researchers; (2) the target population is somewhat familiar with the topics about which they will be questioned; (3) the target population is clearly identifiable and is likely to be responsive to a survey, and; (4) the primary objective is to determine the distribution of characteristics (such as opinions on specific topics or demographic traits) throughout the population. If any one of these elements is lacking, the survey might need to be preceded by qualitative research, or another method may be more appropriate.

Two common methods of generating primary data about people's expectations and values are household and on-site surveys. Household surveys usually quantify how many or what proportion of individuals believe certain things. They typically address data needs at community or coarser scales. Household

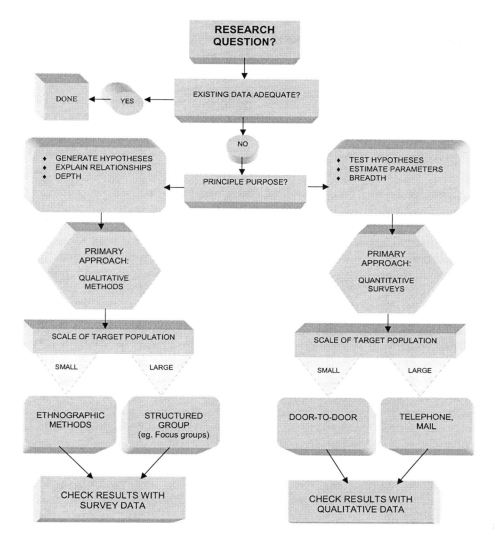

Fig. 1. Decision tree for determining an appropriate social science research method.

surveys can be administered rapidly via telephone, but can be expensive (Fig. 2). Mail surveys are less expensive but take more time. Finally, there is the door-to-door approach, which allows the interviewer to evaluate the quality of data collected and to provide additional observational data. A questionnaire was used to survey attitudes, beliefs, and values for the Interior Columbia River Basin Ecosystem Assessment. The results were used to help identify the implications of alternative management strategies for various groups in the Columbia River Basin. On-site surveys are specialized because they identify persons to be surveyed by their physical presence, activities, and direct involvement at the area of concern.

Numerous factors can make or break a survey. A few of the most prominent factors are: (1) the willingness of local agency personnel to help with logistics and to resolve unanticipated problems; (2) the amount of preparation, such as pretested questionnaires where "glitches" have been worked out so that the respondent, interviewer, data enterer, and analyst all understand the meaning and flow of the questions; (3) the amount of training for interviewers on all aspects of the survey; (4) the length of survey questions — a long a list of questions can turn off respondents so trim unnecessary questions from the survey from the start; and (5) the expertise of analysts who are interpreting the responses.

Qualitative Research. Not all questions lend themselves to survey research. In some cases, it's not possible to design a meaningful questionnaire because little is known about the target population, or about the issues of interest. In other cases, the issues are too complex, personal, or sensitive to be covered in the impersonal format of a standardized questionnaire. Finally, researchers are sometimes more interested in understanding the reasons that motivate certain behaviors, or the relationships between different phenomena, than they are in the proportions of the population subscribing to a particular view. In cases such as these, one or more forms of qualitative field research may be called for. These can be divided into ethnographic and structured group methods.

Ethnographic research seeks to understand the underlying motivations, causes, and essential relationships that account for human behavior and values. Ethnographic research may involve long-term immersion in a culture or community, or it may limited to applied research on one or a few specific problems. Structured group methods are often used to identify groups of key individuals and involve them in structured exercises designed to elicit their views on topics of interest. Focus groups, for example, are a form of structured group methods that can be used to understand the motivations of certain types of people.

Qualitative research also has limitations. Qualitative research is not particularly useful when large groups need to be characterized, such as residents of an ecoregion or a state. Qualitative data are specific to particular individuals and are not meant to describe populations statistically. Natural resource professionals frequently use qualitative approaches to gather information. Such information, however, is often collected and used casually without a plan for systematically processing the information. As a result, this information is not stored for later use and is often of no help to other inquiries. Qualitative research, like quantitative research, should be designed carefully so the information can be used later, or by other researchers.

A simple decision tree can help the researcher or natural resource manager determine which methods are most appropriate for their inquiry (Fig. 1). A brief summary of the steps includes:

1. *Begin by formulating a clear statement of the research problem or question.* This essential first step is often more difficult than one might suppose, but once it is completed, the remaining steps are far more easily accomplished. An impartial third party, a trained social scientist for example, can help define the central problem in ways that can be effectively and efficiently researched.

2. *Evaluate the adequacy of existing information (secondary data) that might be relevant to the research question.*

3. *If new information is needed, determine the primary purpose of the research effort.* Questions to consider include: Is the purpose to explore a largely unknown population, to generate hypotheses, or to deepen understanding of a particular phenomenon? Or is the purpose to test hypotheses already generated, or to estimate population parameters? What is the desired final product — a statistical description, an explanatory account, or a conceptual model?

4. *Consider the size of the population of interest and the size of the sample to be studied.* Social research may focus on individuals, communities, watersheds, States, regions, Nations, or even larger scales. The choice of scale depends upon the research objective (what do we want to learn?), the desired level

Fig. 2. Survey research can generate accurate information on views held by large groups. Such methods, however, are expensive. (Photo courtesy of U.S.D.A. Forest Service.)

of generalizeability (about whom do we wish to learn?), and, of course, upon the resources of time, money, and expertise available for research (what can we afford?).

5. *Cross-check results with data obtained from other methods.*

SELECTED READING

Bleiker, H. and A. Bleiker. 1990. *Citizen Participation Handbook.* Institute for Participatory Management and Planning, Monterey, CA.

Bliss, J.C., and A.J. Martin. 1989. Identifying NIPF management motivations with qualitative methods. *Forest Science* 35(2): 601–622.

Brunson, M.W., D.T. Yarrow, S.D. Roberts, D.C. Guyn, Jr., and M.R. Kuhns. 1996. Nonindustrial private forest owners and ecosystem management: can they work together? *Journal of Forestry* 94(6): 14–21.

Carroll, M.S. and R.G. Lee. 1990. Occupational Community and Identity Among Pacific Northwestern Loggers: Implications for Adapting to Economic Changes, pp. 141–156. In: R.G. Lee, D.R. Field, and W.R. Burch, Jr. (eds.), *Community and Forestry: Continuities in the Sociology of Natural Resources.* Westview Press, Boulder, CO.

Dillman, D.A. 1978. *Mail and Telephone Surveys: The Total Design Method.* Wiley, New York.

Fetterman, D.M. 1989. *Ethnography Step by Step.* Applied Social Research Methods Series, Vol. 17. Sage Publications, Inc., Newbury Park, CA.

Fitchen, J.M. 1990. How do you know what to ask if you haven't listened first: using anthropological methods to prepare for survey research. *The Rural Sociologist* (Spring): 15–22.

Gaventa, J. 1993. The powerful, the powerless, and the experts: knowledge struggles in an information age, pp. 21–40. In: P. Park, M. Brydon-Miller, B. Hall, and T. Jackson (eds.), *Voices of Change: Participatory Research in the United States and Canada.* Bergin & Garvey, Westport, CT.

Kempton, W., J.S. Boster, and J.A. Hartley. 1995. *Environmental Values in American Culture.* MIT Press, Cambridge, MA.

Krueger, R.A. 1994. *Focus Groups: A Practical Guide for Applied Research.* 2nd ed. Sage Publications, Inc., Thousand Oaks, CA.

Overdevest, C. and D.B.K. English. 1994. Understanding People and Natural Resource Relationships: Ouachita National Forest Timber Purchasers and Changing Timber Harvest Policy. In: Ecosystem Management Research in the Ouachita Mountains: Pretreatment Conditions and Preliminary Findings. General Technical Report SO-112. U.S. Department of Agriculture, Forest Service, Southern Forest Experiment Station, New Orleans, LA.

Park, P., M. Brydon-Miller, B. Hall, and T. Jackson (eds.). 1993. *Voices of Change: participatory research in the United States and Canada.* Bergin & Garvey, Westport, CT.

Raval, S.R. 1994. Wheel of life: perceptions and concerns of the resident peoples for Gir National Park in India. *Society and Natural Resources* 7: 305–320.

Whyte, W.F. 1984. *Learning from the Field: A Guide from Experience.* Sage Publications, Newbury Park, CA.

The Evolution of Public Agency Beliefs and Behavior

Why Evolving Agency Beliefs and Behavior Are Important To Ecological Stewardship

Western world conservation movements were forged in the transition and turmoil of rural, agricultural societies becoming urbanized, industrial states during the late 19th and early 20th centuries. So too were the social roles, core beliefs, and management practices of science-based, natural resource professionals. The Western world is once again experiencing similar turmoil at the close of this century as it evolves into an urban post-industrial society linked to economies and cultures around the world. The changing technological, economic, political, and environmental forces driving this post-industrial transition are impacting public and private resource managers everywhere. As with the Industrial Revolution, ecological, economic, and political systems are all swept up in the transition.

This summary was drafted by Robert Szaro with contributions from James J. Kennedy. It is based on the following chapter in Ecological Stewardship: A Common Reference for Ecosystem Management, Vol. III:

Kennedy, J.J. and M.P. Dombeck. "The Evolution of Public Agency Beliefs and Behaviour Toward Ecosystem-Based Stewardship."

KEYWORDS: Policy, bureaucracies, professionals, citizens, mechanistic models, organic models

Traditional assumptions and beliefs about public land management and managers are being challenged once again. No longer is wildfire an unmitigated evil. Foresters and other experts no longer have hegemony over resource management. More inclusive and enduring concepts for natural resources management will be needed in the 21st century. Adaptive approaches to natural resources management — such as the evolving concepts behind ecosystem management — will be unavoidable. Other sectors including medicine, education, and business are also in similar states of wonder and confusion, hope and fear, adaptation and denial.

One of the reasons so many public natural resource agencies are experiencing difficulty adapting to the advent of the 21st century is they have been so successful in this century. Sustained-yield flows of water, wood, recreation, or wildlife goods and services, managed by specialized and bureaucratized professionals, have led to a long era of power and prestige. As natural resources have become more limited at the end of this century, and as citizens have become more knowledgeable, definitions of success and failure have changed rapidly. The professionals and the agencies that employ them have not kept pace. If these changes are to be assimilated, natural resource agencies and their employees will have to shift their learning, their management practices, and their organizational culture.

KEY FINDINGS

✓ **To succeed in the 21st century, public agencies will likely have to do more than reinvent themselves in new shapes and customer orientations**. Public and private organizations must reach deep inside and far back into Western culture to confront and diminish dysfunctional traditions of (1) fragmented thinking and problem-solving, (2) glorification of competition in work and play, and (3) combative, reactionary reflexes to perceived hostile forces, to allow fundamentally new norms and forms of thinking and behavior to take root.

✓ **One of the reasons natural resource professionals are so vulnerable to social criticism is their often-conflicting roles in multiple-use management**. Resource managers are often charged with being both resource *protectors* and *providers* of goods and services (Fig. 1). In the first half of this century, the protector role often focused on maximum timber site or game population productivity (i.e., maximum flows of goods and services) within sustained-yield constraints. This role is maturing today to incorporate a more humble, respectful management focus on healthy, sustainable ecosystem themselves — as well as the multiple-use output endowments they can bestow to current generations (i.e., provider role).

✓ **Abstract concepts like ecosystem sustainability risk public confusion, alienation, and suspicion**. A dramatic shift from concrete public land management messages (e.g., land of many uses, wildlife for today and tomorrow) to abstractions like ecosystem integrity or sustainability that scientists can barely define, measure, or monitor risks alienating the public. Even if professional consensus is reached on these abstractions, there remains the critical task of translating them and their results into understandable and relatable public information. In many cases, these new abstractions need not be evangelized or emphasized if changes in amount and type of public land use can be explained in more traditional ways (e.g.,

Fig. 1. The image of the manager acting on his own is being replaced by the resource manager as a partner working with public constituencies. (Photo courtesy of USDA Forest Service.)

expanding multiple use management to include urban needs and to better ensure resource sustainability).

✓ **Adapting to sustainable ecosystem management should be considered a long-term, evolutionary path, not a fixed, defined target.** Many ecosystem and professional rewards will come from the journey of discovery rather than an explicit, scheduled arrival — such as putting a person on the moon. Faith, perseverance, and adaptability are required in this journey towards "desired future conditions" (Box 1).

✓ **There are powerful riptides of resistance both in and outside natural resource professions and public agencies to ecosystem-based management.** Such is the definition and dynamics of profound change at both the beginning and end of this century. Yet we see little future in the 21st century for closed (vs. open), narrow (vs. inclusive), short (vs. long)

Box 1
Public natural resource managers and agencies are evolving from:

Cold-war conservation to Collaborative conservation
Patron bureaucracies to Partnership organizations
Line-staff tiers to Open, adaptive teams
Linear-thinking specialists to Synergistic integrators
Output staff managers to Social value managers and stewards
Technical functionalists to Ecosystem-based management facilitators.

---

term, or machine- (vs. organic-) model thinking and behaving in public or private organizations, regardless of how comfortable and secure old ways of thinking might feel.

✓ **Natural resource organizations will require a change of heart and inclusive, integrated, system-oriented thinking to re-establish public trust**. Ecosystem-based or collaborative stewardship is an appropriate organic model of thinking that continues on the evolutionary path promised by multiple-use management (1950s) and envisioned in the 1970s environmental era. This broader, more collaborative and adaptable ecosystem stewardship thinking, when it is equitably and appropriately applied, must include other interrelated economic and sociocultural systems. Ecosystem stewardship thinking is also compatible and on a parallel path to creating "learning organizations."

PRACTICAL CONSIDERATIONS

Natural resource agencies in the first two-thirds of this century focused on holding external "entropy forces" at bay. Resource agencies and their professionals used regulations, technology transfer, use control, road and infrastructure development, strategic planning, and increased technological expertise to maintain technocratic control over natural resources. Patriarchal line-staff hierarchies, tight job classifications and promotion eligibility, and rigid budget accountability were internal devices to combat *internal* "organizational entropy". In the process of battling hostile internal and external forces, resource professionals sometimes alienated themselves from the land, their colleagues, public land users, and other segments of society. It is time to rethink many basic agency assumptions and core beliefs.

The public land manager's role as provider of goods and services has evolved and matured beyond an output quantity focus. Today that role includes more consideration of *outcomes* directly and indirectly generated by outputs over expanded time and space (e.g., timber or grazing outputs that yield employment and community sociocultural outcomes, as well as wildlife habitat change), and focuses on customer or stakeholder *service* (e.g., how output qualities and outcome impacts are evaluated by users and other stakeholders).

The public land manager's provider role will evolve towards greater incorporation of stakeholder service and outcome concepts in a broad social value context. More than public servants, managers today are often social value brokers and conflict management facilitators. Thinking must expand from an ecosystem focus to incorporate regional economic and sociocultural systems. Rather than just *ecosystem managers*, public land managers must focus more broadly on *sustainable systems* management (including ecological, sociocultural, and economic systems) in a more collaborative, partnership relationship.

Ecosystem stewardship or collaborative management is a promising, organic-model orientation toward public lands and waters for the next century. These approaches are being pursued in the United States and Europe, and increasingly in other parts of the world, under different names and forms. Such approaches require considerable shifts in public land time and space relationships with (1) ecological subsystems, (2) stakeholders living and unborn, and (3) managers. Yet in many ways, ecosystem-based stewardship is a logical evolution and a maturation of traditional multiple-use, sustained-yield conservation values and concepts, with its more inclusive concern for multiple-use (and non-use) values and the elevation of long-term site productivity from a constraint to a primary sustainability focus.

SELECTED READING

Council for Environmental Quality (CEQ). 1995. *The Ecosystem Approach: Healthy Ecosystems and Sustainable Economies.* Vol. 1. Washington, DC.

Dombeck, M.P., J.W. Thomas, and C.A. Wood. 1997. Changing Roles and Responsibilities for Federal Land Management Agencies. In: J.E. Williams, C.A. Wood, and M.P. Dombeck (eds.). *Watershed Restoration: Principles and Practices.* American Fisheries Society, Bethesda, MD.

Gunderson, L.H., C.S. Holling, and S.S. Light (eds.). 1995. *Barriers and Bridges to the Renewal of Ecosystems and Institutions.* Columbia University Press, New York, NY.

Hollick, M. 1993. Self-organizing systems and environmental management. *Ecosystem Management* 17(5): 621–628.

Kennedy, J.J., and J.W. Thomas. 1992. Exit, Voice and Loyalty of Wildlife Biologists in Public Natural Resource/Environmental Agencies. In: W.R. Mangum (ed.), *American Fish and Wildlife Policy: The Human Dimension.* Southern Illinois Press, Carbondale, IL.

Knight, R.L., and S.F. Bates (eds.). 1995. *A New Century for Natural Resource Management.* Island Press, Washington, DC.

Senge, P.M. 1990. *The Fifth Discipline.* Doubleday, New York.

Schiff, A.L. 1966. Innovation and administrative decision making: the conservation of land resources. *Administrative Sciences Quarterly* 11(1): 1–30.

Wheatley, M.J. 1992. *Leadership and the New Science.* Berrett-Koeher, San Francisco, CA.

Wiersum, K.F. 1995. 200 Years of sustainability in forestry: lessons from history. *Environmental Management* 19(3): 321–329.

Processes For Collaboration

Why Collaborative Processes Are Important To Ecological Stewardship

Natural resource management issues are more controversial than ever, with more varied interests and more people capable of bringing the process to a halt. In many cases, national legislation, legal solutions, and special interest groups prevent important concerns from being integrated into the process. Adversarial approaches often prevent the consideration of different world views, impede the use of systems thinking, overlook the incorporation of social needs and local engagement, and compromise ecosystem sustainability. In addition, portions of the public are losing trust in government and traditional sources of authority to make comprehensive judgements and decisions about natural resources. In reaction to conflict and legal paralysis, agencies, organizations, and citizens around the United States are exploring how collaborative processes can identify common ground to resolve, or at least to manage, differences.

This summary was drafted by Andrew Malk with contributions from Mark Hummel, Jeff Romm, Carl Reidel, Elaine Hallmark, Gordon Brown, and Nels Johnson. It is based on the following chapter in Ecological Stewardship: A Common Reference for Ecosystem Management, Vol. III:

Hummel, M., B. Freet, S. Mills, and M. Phelps "Collaborative Processes for Improving Land Stewardship."

KEYWORDS: Consensus, controversy, information sharing, participation

---

Box 1
Information-sharing to Build Trust in Southeast Alaska

A U.S. Forest Service team in Wrangell, Alaska, adopted a collaborative approach to develop an environmental assessment for the management of a wildlife observatory. The observatory is a favorite destination for outfitters and guides to take their clients, because black and brown bears are frequently seen feeding on the large runs of pink salmon (Fig. 1). However, rapidly growing use of the observatory threatened to chase away the bears, increase the risk of injuries to people and bears, and diminish the visitors' natural experience. Instead of simply referring he problem to a group of specialists, the team developed a series of partnerships. They involved outfitters and guides, local residents, Native Americans, and the Alaska Department of Fish and Game to help define the problem. Researchers from the Alaska Department of Fish and Game and Utah State University took the lead in collaring bears and plotting home ranges. Once collected, the research results were distributed to each of the groups. In partnership with these groups, the U.S. Forest Service team identified acceptable management alternatives. This information sharing process demonstrated that the agency was acting with the most recent research on bear-human interactions, and the final decisions to limit visitation were based on information that all had access to. Despite the potential for serious disagreements over the decision, there was no administrative or legal appeal — a rarity in these adversarial times — because all stakeholders were able to anticipate the results of the research.

Collaboration means working together effectively on issues, problems, and situations of mutual interest. In some cases, the process begins when situations with conflict and controversy identify the need for the various parties to work together. In others, organizations recognize that they do not have the resources or capacity to implement independently ecosystem approaches to natural resources management. In collaborative processes, people work together by sharing information, identifying mutual interests, and building understanding and trust. Collaboration may include information-sharing processes, partnerships, and agreement-seeking processes. These approaches are not exclusive. For example, a process could easily involve information sharing and partnerships. All three processes can be oriented toward mutual learning and group investment in the problem-solving process (Box 1).

Fig. 1. An Alaskan brown bear in southeastern Alaska feeding on runs of salmon. (Photo by Dave Menke, U.S. Fish and Wildlife Service.)

Collaboration creates opportunities to assume shared responsibilities for sustaining natural and social systems. It can also provide opportunities to build durable partnerships that can make agency managers and community leaders more effective at meeting their responsibilities and getting the job done. In addition, collaboration creates opportunities to reduce or resolve conflicts in potentially less formal, less costly ways than legislation or adversarial legal proceedings. Collaborative processes provide opportunities to form innovative partnerships that can respond to the complexity of environmental, social, and economic conditions more effectively than resource agencies can on their own.

KEY FINDINGS

✓ **Collaboration comes in different forms**. It can be initiated from a resource management agency, or it can start within a community or private organization. Collaboration may be limited to sharing information or it can seek to establish formal and binding agreements (Box 2).

✓ **Collaboration in ecosystem management requires learning about social values, differing world views, and the ecosystems themselves**. Effective collaboration is rarely engineered. It often comes about through intuitive learning about a community, its history, and leaders. Since collaborative processes are as much art as they are science, there is a continuous need for listening and observing them in action. Through periodic reassessment, participants can learn what does and does not work in a particular community or setting. Collaborative processes should be adapted on the basis of this learning.

✓ **Collaborative learning encourages people to think "differently" about controversies and policy decision situations**. Thinking differently involves reframing debates — literally changing the language and perceptions of natural resource conflicts. In particular, collaborative learning:

- Stresses improvement rather than solution,
- Emphasizes situation rather than problem or conflict,
- Focuses on concerns and feasible changes rather than desired future condition,
- Targets progress rather than success, and
- Encourages systems thinking rather than linear thinking.

✓ **The world around the collaborative process — as well as the people and organizations participating in it — will change rapidly**. Change is a critical dimension of collaborative processes since different participants will have different stakes in various economic, social, and political developments — some of which may change on a nearly daily basis. Collaborative processes, therefore, need to respond to such changes as market structure and performance, technologies, economic and demographic trends, and environmental quality. Adaptive management techniques that monitor and incorporate key trends relevant to the goals of the collaborative process can strengthen the collaboration. Such techniques may include periodic joint assessments of project progress, participatory monitoring of key environmental, social, and economic trends, and the use of participatory appraisal methods.

> **Box 2**
> **Categories of**
> **Collaboration**
>
> *Agency-Based or Community-Based?* Agency-based efforts usually include land management plans, landscape assessments or designs, project plans, or putting the plans into action.
>
> *Community-based* efforts may involve a planning and zoning department developing a comprehensive community plan or ad hoc organizations forming to address community issues.
>
> *Information-Sharing or Agreement-Seeking? Information sharing* involves presenting ideas to a larger group and asking for feedback and additional ideas. The objective is to understand the concerns and ideas of others and integrate them into a better project or solution.
>
> *Agreement seeking* involves trying to agree as a group on recommendations to a decision-maker or agreement on the decision itself. The group itself integrates members' concerns and ideas into a better recommendation or decision.

✓ **Many process options and training courses are available, but the key ingredient is willingness.** This means a willingness to talk, listen, explore, learn, innovate, propose, refine, and act on ideas; a willingness to let go of control in order to make progress; a willingness to share responsibility with stakeholders for fulfilling agency mandates. One should start wherever in the process that seems to make sense. In some cases that may mean building relationships before being able to identify a situation or issue to resolve. In others cases, a partnership idea may be useful as a way of building relationships.

✓ **A negotiated agreement is only the beginning.** Once an agreement is reached, implementation can take considerable time and collaborative effort, sometimes more time than the agreement itself. The reality of making the agreement happen often involves more detail than the original agreement. In addition, the individuals implementing the agreement are sometimes different than those who negotiated the agreement. The process still requires considerable effort even after participants forge an agreement.

✓ **Experimental science can be used in conjunction with collaborative processes to facilitate systematic learning within organizations and communities.** This provides tangible opportunities for shared development of hypotheses, implementation of activities, and monitoring and interpretation of results. Scientific findings can also reduce the difficulties of attaining agreement by demonstrating the improbability of extreme expectations that would otherwise tear people apart.

PRACTICAL CONSIDERATIONS

Managers should consider several key questions when embarking on a collaborative process. These include:

How do we select the approach that works for our situation? In developing collaborative relationships, it is important to clarify whether you are engaging in a form of information sharing or in an agreement-seeking process and whether the process is agency-oriented or community-based. Agency-oriented information sharing processes have evolved a long way from the typical public meeting of the past. Examples of new modes or models are open house sessions, open space meeting, search conferencing, and participative design workshops and some advisory groups (Fig. 2). Agreement-seeking processes include partnering, negotiation, mediation, facilitated consensus process, which might also be an advisory group, and negotiated rulemaking.

How do I decide when and where to work for a consensus decision? In what circumstances is a consensus-building process most likely to succeed? Consensus on recommendations or a decision is often the target of agreement-seeking negotiations. Consensus processes are most often effective when the following factors exist: the parties in dispute truly want a settlement; the parties are in an ongoing relationship with one another; the parties are willing to have a third, neutral party facilitate; and the parties have external reasons to want a settlement, such as time or money pressures. The advantages of consensus processes are often better known than the disadvantages and include decisions being made at the level at which the agreement is implemented, rather than at the level of national politics or federal courts. As a result, participants often display greater acceptance of the solution. However, notable disadvantages exist. In an agreement-seeking process, every individual is

Fig. 2. Information sharing meetings are typically more participatory today than public meetings of the past. (Photo courtesy of USDA Forest Service.)

granted veto power. As a result, outcomes are sometimes satisfactory only at the lowest common denominator. Consensus processes generally require considerable time and cost. If feelings and wishes are not combined with facts, social pressure may push the agreement-seeking group to conform with the majority opinion.

How do I identify who should be involved in a collaborative process? First decide which kind of process (see above). Also decide whether you are trying to generate interest, whether you want everyone with an interest to participate or whether you want a smaller group of representatives. The key is to be inclusive in the way that is appropriate to the forum you are using. The assessment phase of a consensus process helps to identify and reach an agreement on the appropriate participants.

How do I deal with situations where one or more parties do not trust each other enough to try working together? How do I deal with fringe groups and manage dissent? Many important natural resources decisions involving the public must bring opposing groups such as loggers and environmentalists together (Fig. 3). Providing a forum in which parties can hear each other's interests can help broaden their views and identify common values. It may also help to talk about desirable and feasible change rather than desired outcomes, since desired outcomes may seem overwhelming. By emphasizing a larger view of the situation that incorporates the perspectives of the participants, you help build a trust among participants that you will not ignore particular interests. Some groups create conflict, because they are not being heard. By giving them your ear, a seat at the table, and a voice in your management, you are giving them the influence they want. This situation encourages them to move from simply saying "no" to working on real solutions.

What incentives can I use to foster participation? How do I evaluate trade-offs associated with incentives? It is important to see a potential benefit for participating. Will the outcome truly be affected? Agencies need to clarify how they are going to work with or use the participants' contributions. Interagency collaboration needs to be based upon a clear understanding that it is worth the effort.

Fig. 3. One of the major challenges of ecological stewardship is to find common ground between antagonists. (Photos courtesy of USDA Forest Service.)

SELECTED READING

Cormick, G.W., et al. 1996. *Building Consensus for a Sustainable Future: Putting Principles into Practice.* Ontario, Canada: National Round Table on the Environment and the Economy.

Daniels, S.E. and G.B. Walker. 1996. *Collaborative Learning: Improving Public Deliberation in Ecosystem-Based Management.* Oregon State University, Corvallis, OR.

Fisher, R., and W. Ury. 1981. *Getting to Yes: Negotiating Agreement Without Giving In.* New York: Penguin Books.

McAllister, W.K. and D. Zimet. 1994. Collaborative Planning: Cases in Economic and Community Diversification. Report FS-575. Washington, DC: USDA Forest Service.

McWilliams, R. and F. Patten. 1995. Partnerships for Progress: Forest Service's Collaborative Approach to Sustaining Forests and Rural Communities. Paper presented at annual meeting, Society of American Foresters, Portland, ME.

Robertwhaw, F., et al. 1993. *Conservation Partnerships: A Field Guide to Public-Private Partnering for Natural Resource Conservation.* National Fish and Wildlife Foundation, Washington, DC.

Schindler, B., et al. 1993. Managing Federal Forest: Public Attitudes in Oregon and Nationwide. *Journal of Forestry* 91(7).

Thomas, J.W. 1995. Engaging People In Communities of Interests. Speech presented at C.E. Farnsworth Memorial Lecture, SUNY, Syracuse, NY.

Wondolleck, J.M. and S.L. Yaffee. 1994. *Building Bridges Across Agency Boundaries: In Search of Excellence in the United States Forest Service.* USDA Forest Service research report. Pacific Northwest Research Station, Seattle, WA.

Regional Cooperation

Why Regional Cooperation Is Important To Ecological Stewardship

It is widely recognized that ecosystems, natural communities, and ecological processes are fragmented by political jurisdictions. It is equally important to recognize that key factors shaping natural resources management — information, public values, financial and technical resources, and political and economic power — are also fragmented across geographic boundaries, social groups, organizations, agencies, and disciplines. Ecological stewardship at the landscape level can only be achieved by creating an interactive network of ideas, information, and capabilities. The challenge of a pluralistic society is to make sense of all this diversity, and assemble coalitions of interests and values that enable society to move forward. Hence there is a need to promote interaction across regions for several reasons (Box 1).

This summary was drafted by Andrew Malk with contributions from Steven Yaffee, Robert Dumke, and Hanna J. Cortner. It is based on the following chapters in Ecological Stewardship: A Common Reference for Ecosystem Management, Vol. III:

Yaffee, Steven. "Regional Cooperation: A Strategy for Achieving Ecological Stewardship."

Johnson, Kathleen M., Al Abee, Gerry Alcock, David Behler, Brian Culhane, Ken Holtje, Don Howlett, George Martinez, and Kathleen Picarelli. "Management Perspectives on Regional Cooperation."

KEYWORDS: Fragmentation, partnerships, stakeholder, bureaucracies, collaboration

---

Box 1
Five Practical Reasons for Regional Cooperation

Cooperation between agencies and groups in natural resources management is never easy, but it can be essential for ecosystem approaches for the following reasons:

1. *Acquiring necessary information*: Because a single agency seldom controls all information relevant to the management of the landscape, effective management requires better sharing of information across disciplines;

2. *Accumulating and sharing resources*: Regional cooperation can improve the efficiency and cost-effectiveness of resource management, and has the potential to provide landowners and businesses with a measure of certainty about what to expect. Shrinking public budgets are making public-private partnerships necessary to increase access to labor, expertise, authority, funds, and equipment. In some rural resource-dependent areas, public agencies are among the few sources of expertise needed to achieve community objectives;

3. *Making effective decisions that stick*: Today, a wider range of groups is interested in and affected by public resource management. Determining appropriate courses of action, and implementing them over time, require effective interaction among a diversity of groups and interests. Decisions made arbitrarily are likely to be challenged in administrative or judicial proceedings;

4. *Building support, ownership, and a sense of civic responsibility*: The best collective choices are those that are supported by a broad set of interests who feel ownership in the solutions to problems being addressed. Without this support, groups will resist taking appropriate action, and public agencies will not be able to carry out ecological stewardship actions, and;

5. *Influencing knowledge and values*: Ecologically responsible resource management will only succeed over the long term if it moves forward hand-in-hand with an effort to inform both the public and agency staff, and influence their values. Cooperative relationships can provide a structure that promotes education and refinement of values.

Cooperation is an innocuous term, but by all accounts it is problematic. Parents implore their children to cooperate, many institutional studies conclude that more interagency cooperation is needed, and no one is opposed to cooperation as a general concept. Like motherhood and apple pie, cooperation is a long-standing, shared American value that is at least as strong a component of American history as competition. Concepts of cooperation evoke images of community and neighborhood as well as the underlying rationale for a system of democratic government, including concepts of federalism, pluralism, and representation. But the simple fact that cooperation is invoked prescriptively so often in sites ranging from a schoolyard to the U.S. Congress suggests that achieving effective cooperation between individuals and institutions is not so simple. As resource management moves towards ecosystem-based approaches, cooperation and collaboration at appropriate geographic scales becomes critical.

KEY FINDINGS

There is no one right way to stimulate cooperative interactions, nor is there a single geographic or functional scale at which cooperation makes sense. Some approaches focus on specific problems, while others allow emerging issues and concerns to be aired. Some are *ad-hoc* and temporary, others are institutionalized and longstanding. Where the problems require a sustained level of interaction, structures need to be created to foster and manage the interactions. Flexibility is important, however.

Effective management requires better sharing of information across disciplines within agencies, more effective interagency relationships, and development of multiparty networks among public and private individuals and groups. No longer do natural resource managers exclusively have the "right answers" to management, nor should they take on the burden of finding them on their own. Cooperation is not a stand-alone task that should be put in a box called public involvement. Instead, cooperation should be viewed as a critical element of how one goes about doing many other things.

Although most experience with ecosystem approaches is recent, important lessons can be drawn from case studies and surveys of existing efforts. Factors that facilitate cooperation at a regional level can be grouped into three areas:

✓ **Situational Factors.** The primary challenge for managers facing regional management "opportunities" is to determine how to take advantage of factors inherent in their situations that will positively influence the outcome. Having a shared sense of a problem or threat to an area, or building on a feeling that an area or region is unique, sometimes motivates diverse individuals to overcome prior conflicts and work together. Sometimes, groups are forced to look for objectives above the current conflict to find shared goals. That is, what starts as a battle over a specific proposed action evolves into a broader look at a community's future. A sense of place is also helpful in fostering a cooperative spirit. Prior relationships between individuals and organizations often facilitate regional cooperation. For example, a network of local ranchers and residents along a stretch of the Mexican border in New Mexico and Arizona with long-standing relationships as neighbors was key to the formation of the Malpai Borderlands Group. Its twin goals are to maintain ranching as a viable economic activity in the face of mounting development pressures, and to restore fire's important role in the ecosystem to counter declining range productivity and wildlife habitat loss (Fig. 1).

✓ **The Processes Used**. The perception of fairness and equitable treatment of different groups' concerns is important for cooperative efforts to succeed. At a minimum, all relevant stakeholders need to be involved in a significant way from the outset of the project. Whether the project will be fruitful will

Fig. 1. Bill MacDonald, a local rancher, helped catalyze the Malpai Borderlands Project. (Photo courtesy of The Nature Conservancy.)

depend in part on its perceived legitimacy. Process management skills are also important. Many fledgling efforts die because of poor interpersonal dynamics. Indeed, many natural resource managers sought their careers to work "in the woods" in part to avoid having to deal with other people. Having individuals who can mediate between conflicting positions and combative personalities has been a key to success in many efforts, yet few resource managers have been trained to manage or facilitate participatory processes. Finally, the presence of dedicated leaders — whether agency resource managers or field staff, community leaders, landowners, or elected officials — is indispensable. As one respondent to a University of Michigan survey on ecosystem management projects said, "It always boils down to key talented people who are willing to invest themselves over and above the call of duty."

✓ **The Institutional Context**. The presence of one or more organizations with technical, human, and financial resources needed for planning and implementation is usually key to a successful regional cooperative effort (Fig. 2). The mere presence of such institutions, however, is often not sufficient for success. Participants must believe that agency and elected officials support the cooperative process, and that they will be likely to follow through on agreements reached through it. It is devastating to all concerned when stakeholders invest a lot of time and energy to find a reasonable course of action that all can live with only to see it disregarded by agency or elected officials. Agency personnel participating in regional cooperative efforts should be assured of clear support from higher line staff and supervisors. Other institutional factors can also be key to the success of regional cooperation. Gaining support from community leaders and hiring project personnel from within the community are often cited as factors in the success of ecosystem management projects. Sometimes creating opportunities for interaction is enough; at other times, structured incentives for action are needed. Such incentives may require the "carrots" of seed monies or technical assistance, or a regulatory outcome that takes effect if a cooperative effort does not succeed. But they may also be simply providing individuals with flexibility they previously lacked, or certainty that if they "do the right thing," they will not be penalized for doing so.

Fig. 2. Tackling planning and management issues often requires the resources, expertise, and information of more than one agency. (Photo courtesy of USDA Forest Service.)

Laws, political realities, financial necessities, and other factors influence the decision of an organization or group whether to participate in a cooperative effort. The potential for "making a difference" or "having an impact" motivates stakeholders to be involved. Projects that are rooted in the local community are better received than those perceived as top-down agency directives or outsider initiatives.

PRACTICAL CONSIDERATIONS

Case studies show that a variety of factors can impede regional-scale cooperative efforts. These factors include limited resources, inadequate interpersonal skills, and ineffective process management. A wide range of attitudinal factors constrain effective interaction, including conflicting goals, values, and missions, public opposition and fears, organizational turf, and intergroup attitudes. Many regional institutions have been ineffective because they are dominated by the parochial concerns of their members and are unable to rise above them. The following factors often impede regional cooperation.

Attitudes. Public skepticism and outright opposition are often cited as among the most difficult barriers to the development of collaborative ecosystem management projects. Opposition is often rooted in misperceptions about what ecosystem management entails (e.g., focused only on preservation, or managed only by scientists) and longstanding mistrust of government. Often these attitudes have been reinforced by a history of polarized confrontations over natural resources management that exaggerate differences between various stakeholder groups. In reality, there is often more common ground than antagonists realize. Roundtables to discuss problems, joint information gathering, and other strategies can be a starting point to change attitudes.

Lack of Process Management Skills. Cooperative efforts depend on processes that require managing complex tasks. Few natural resource managers were trained to handle such tasks as establishing ground rules, managing data, creating a welcome and open environment for all stakeholders, facilitating the participation of diverse personalities, and mediating conflict.

Differences in Data Collection and Analysis. Private firms and public agencies frequently collect technical information in different ways and analyze it for different objectives. Without agreement on the credibility of data and which data to use, regional efforts can flounder in the identification of problems and alternative management options — vital steps in a successful regional project. Different ecological and landscape classification systems used by various federal and state agencies, non-governmental organizations and corporate landowners are one example of this problem.

Shortages of Funding, Trained Personnel, and Inflexible Budgeting and Administrative Procedures. While cooperative efforts can be seen by public resource managers as one way to cope with fiscal shortages (by leveraging the shared resources across multiple stakeholders), they are typically among the first tasks to be cut when budgets tighten. Organizations tend to retreat to their core activities, and the time and effort needed to build relationships outward are often seen as nonessential. Even when funding is available, the way organizations account for and allocate funding tends to hamper cooperative ecosystem management efforts. Lack of flexibility in procedures for implementing agreements also hampers cooperative efforts.

Restrictive Policies. Various federal and state policies can inhibit voluntary cooperation to achieve ecosystem stewardship goals. For example, the Federal Advisory Committee Act (FACA) has been a significant barrier to communication and coordination among federal agencies and other stakeholders. Fear of FACA-inspired lawsuits has hampered a number of creative cooperative efforts, such as the Applegate Partnership in Southwestern Oregon. Corporations may decide not to cooperate on the basis of litigation concerns prompted by the Sherman Anti-Trust Act and the Freedom of Information Act. Finally, shifting political priorities in Congress and executive agencies can leave cooperative efforts without money and agency support.

People cannot be forced to cooperate. Resource managers, however, can create the conditions under which people who would benefit from cooperative interactions can find ways to realize those benefits. There are many ways to structure cooperative approaches, ranging in levels of structure and institutionalization from computer conferences to ad hoc task forces to federal grazing boards. Two images — knowledge pools and relationsheds — may be helpful in conceptualizing ecosystem-level approaches to building information and relationships across space and time. Cooperation involves real costs of time and other resources, and hence needs to be approached in a way that acknowledges and minimizes these costs. Most resource management personnel were not selected on the basis of interpersonal skills, but cooperative working arrangements clearly require them. Such skills are even more important because of a considerable amount of public opposition and skepticism about government and ecosystem approaches. Still, early experience from collaborative approaches in ecosystem management is promising. Individuals have been able to overcome many of the obstacles and accomplish specific objectives, ranging from improved communication to restoration of ecological processes.

SELECTED READING

Keystone Center, 1996. *The Keystone National Policy Dialogue on Ecosystem Management*. The Keystone Center, Boulder, CO.

Knight, R.L., and P. Landres. 1998. *Stewardship Across Boundaries*. Island Press, Washington, DC.

Yaffee, S.L., A. Phillips, I. Frentz, P. Hardy, S. Maleki, and B. Thorpe. 1996. *Ecosystem Management in the United States: An Assessment of Current Experience*. Island Press, Washington, DC.

Yaffee, S.L., and J.M. Wondolleck. 1996. Building Bridges Across Agency Boundaries. In: K. Kohm and J. Franklin (eds.), *Creating a Forestry for the Twenty-First Century: The Science of Ecosystem Management*. Island Press, Washington, DC.

Social and Cultural Dimensions

♦ *Cultural and Social Diversity and Resource Use*

♦ *Social Classification*

♦ *Social Processes*

Cultural/Social Diversity and Resource Use

Why Cultural and Social Diversity Are Important To Ecological Stewardship

Racially, culturally, and socially, the United States is one of the most diverse countries in the world. This diversity has increased steadily since American independence and is projected to grow rapidly in coming decades. Nationwide, the U.S. Census Bureau classifies just over 25 percent of the population as belonging to "minority" groups. In California, there is now no dominant majority population. But racial diversity is only the simplest representation of America's cultural diversity. Each of the Census groups is composed of people with a variety of cultural and social traditions and practices. For example, while Latinos may be united by a shared language, they have diverse national and cultural origins, varied economic interests, and different social institutions. And, of course, the white "majority" population in the United States is itself extremely diverse socially and culturally. Cultural and social diversity is therefore both a pronounced and dynamic feature of American society (Box 1).

This summary was drafted by Nels Johnson with contributions from Carol Raish, William deBuys, Susan Lees, Hanna Cortner, Bill Sexton, and Alfonso Peter Castro. It is based on the following chapters in Ecological Stewardship: A Common Reference For Ecosystem Management, Vol. III:

deBuys, William, Muriel Crespi, Susan Lees, Denise Meridith, and Ted Strong. "Cultural and Social Diversity and Resource Use."

Raish, Carol, Lynn Engdahl, William Anderson, Donald Carpenter, Muriel Crespi, Paul Johnson, Les McConnell, and Earl Neller. "Resource Management Strategies for Working with Social and Cultural Diversity."

KEYWORDS: ethnic diversity, local knowledge, stereotypes, cultural practices

---

Box 1
Cultures As Dynamic Systems

No culture is frozen in time. Cultures change, and so do the people who give them life. Native and historic American cultures continue to disappear at a rapid rate, even as immigrants introduce new cultures to the United States. Cultural groups may grow larger or smaller and gradually alter their life-styles and environmental practices. The significance of a particular practice, such as hunting or fishing, may change with time. Diversity within a group must also be recognized, for men and women may have different knowledge bases and engage in different practices — differences that may grow or narrow over time. For example, women are much more engaged in outdoor recreational activities today than they were a generation ago. It would be difficult and probably pointless to try to catalog each of the cultures in the U.S. and monitor the changes each exhibits. What matters for natural resource managers is that the diversity of cultural groups in a particular place and at a particular time will influence how natural resources are viewed and used. Failure to recognize the current state of cultural diversity can lead to missed opportunities to build widely supported and responsive ecosystem management programs.

Cultural and social traditions and beliefs shape resource use and environmental views in important ways, for human and "natural" systems are deeply intertwined. Moreover, given the dominance of human influence throughout most of the natural world, the majority of natural systems depend for their survival on some degree of sympathetic human stewardship. If natural resource management programs are to win popular support in the United States today, they must integrate different social and cultural perspectives in their planning and implementation. On the other hand, if people do not "buy in" to management programs, they will be unlikely to abide by them. Through their day-to-day actions in the landscape, or through the politicians they elect, people will defeat systems in which they do not believe.

There are also important moral, spiritual, legal, and political reasons to address cultural and social diversity in natural resource management. From a moral point of view, addressing social and cultural diversity is necessary if society is to provide environmental justice and protection for all citizens. From a spiritual point of view, many people, especially Native Americans, hold certain lands and waters sacred — beliefs and values that are often overlooked in economic and scientific approaches to natural resource management decisions. From a legal and regulatory point of view, consultation with local communities may be obligatory under various state and federal laws such as the National Environmental Policy Act (NEPA), the American Indian Religious Freedom Act (AIRFA) and the National Historic Preservation Act (NHPA) (Fig. 1). Finally, from a political point of view, the fair integration of multiple points of view and diverse resource practices in ecosystem management contributes to the health and vigor of American democracy.

KEY FINDINGS

Public agencies need to recognize several core issues if they are to account for cultural and social diversity, in sustainable and equitable resource management. These include:

✓ **The importance of human uses of certain plants, animals, and places to the preservation of cultural traditions**. People use landscapes in myriad ways, and some uses — even those that seem trivial to outsiders — have great

Fig. 1. A resource manager consults with members of a local Native American community. (Photo courtesy of USDA Forest Service.)

cultural significance. The collection of herbs, clays, or plant fibers, or of foods, minerals, or building materials; the right to hunt, fish, graze livestock, or capture animals of ritual significance; the right of access for travel, retreat, or pilgrimage; all of these "uses" may embody aspects of a relationship to a place, to an ecosystem, that co-evolved with the identity of the people who engage in them. Even when discontinuance of a use produces no immediate economic hardship, it may cause other kinds of injuries that are no less serious for being intangible.

✓ **The potential contribution of local cultural knowledge and practices to sustainable resource management**. Resource managers are increasingly realizing the value of accumulated local cultural knowledge. For example, the Forest Service, Department of Interior, many state natural resource management agencies, and The Nature Conservancy have accepted the use of "prescribed fire" in ways that resemble the burning practices of some Native Americans. Prescribed fire is now viewed in many areas as an important tool for restoring ecological health.

✓ **The possible friction between cultural groups when their understandings, values, interests and practices conflict**. Two or more groups, viewing the same landscape, may reach profoundly different conclusions about what the landscape *means* and what its appropriate uses might be (Fig. 2). Even when decisionmakers have access to high quality cultural information, and reach decisions that address the concerns of the affected groups, conflict may nevertheless result. Understanding differences among groups and maintaining a constructive dialogue with each of them, however, can minimize cultural friction over resource management.

✓ **The difficulty of achieving satisfactory cross-cultural agreement when competing groups have different levels of economic and political power**. Consensus on the proper use and disposition of public lands can be difficult, even when competing groups have equal economic and political footing and access to decision making. When competing groups differ significantly in their economic and political power or in their acceptance of decision making processes, the difficulty of fairly balancing the interests of all cultural groups in a given ecosystem increases markedly. Public agencies

Fig. 2. Devils Tower National Monument is a sacred site to Native Americans as well as a popular destination for rock climbers. Differences in the perception of the monument as a sacred place versus a place to climb has created tensions between its users. (Photo courtesy of Ross Harpestad and R. Scott Hawkins.)

may not be able to address political and economic imbalances, but they are responsible for ensuring that all citizens — regardless of economic or political standing — have access to planning and decision-making processes.

✓ **The tendency — perhaps universal among humans — to stereotype people from groups other than their own**. Group stereotyping can lead to serious errors of perception and understanding. Not all African-Americans reside in the "inner city", and not all American Indians live on reservations. Rural African-Americans and urban Indians have different knowledge bases and interests than their respective urban and rural counterparts, even while they share many interests and values. It is rarely wise to generalize from one group to another — to assume, for example, that a description of the land-use practices of one Native American group applies to those of others. It is essential to guard against the use of stereotypes and to understand each group, and each interacting set of groups, as a unique expression of a specific place and time.

✓ **The distinctiveness of bureaucratic cultures, including those of natural resource management agencies**. Institutions create their own culture. They develop special vocabularies and terms of expression that are meaningful to personnel within the agency but may be confusing or unintelligible to the general public, or even to personnel representing similar organizations. The values, habits, and traditions that develop within resource management agencies help shape the interactions of their employees with the outside world. Agency interactions with their stakeholders can be understood as intercultural relationships. As such, these interactions can be more productive when agencies recognize these differences, explain their agency's values and traditions to others, and communicate clearly with those outside the agency's culture.

PRACTICAL CONSIDERATIONS

The biggest challenge for resource managers is to engage diverse groups directly and meaningfully in planning, decision making, and implementation. Whether the engagement is through a planning process, advisory committee, or a project partnership, any consultative process to integrate diverse cultural groups into natural resource management should strive to incorporate the following tasks:

Identify socioculturally diverse groups. Land management agencies must be able to identify and understand the diverse groups who use public lands and may be affected by agency projects (Fig. 3). Managers learn to identify and understand the cultural values, perspectives, and resource use practices of these groups in many ways. Some are directly related to their roles as leaders and facilitators of the public participation process. From reading published sources on the area to in-depth interviews with agency sociocultural specialists and discussions with employees who represent different groups, managers grow to know and understand the people and resources of the land they administer. They may also initiate and participate in ethnographic research or community studies that provide baseline information on resource use and cultural practices.

Communicate with socioculturally diverse groups. Communication between public agencies and affected groups is critical to project planning and implementation, but sometimes breaks down because of mutual misunderstanding. Many problems stem from the ways agencies communicate and "do business." The bureaucratic culture itself, limitations of the public participation process, lack of sufficient inter-cultural knowledge, and inappropriate communication techniques all hamper cross-cultural communication and understanding. Some agencies are now using early contact with interested groups, personal face-to-face interaction, meetings that are tailored to individual group needs, and continuing long-term contact and education programs to improve cross-cultural communication (Box 2).

Incorporate information from socioculturally diverse groups into land management planning. Once sound communication is established, the most challenging part of working with culturally diverse groups may be using their information in a

Fig. 3. Inner city residents are important users of public lands and waterways surrounding metropolitan areas. (Photo courtesy of USDA Forest Service.)

> **Box 2. Engaging the Hispanic Community in the Santa Fe National Forest, New Mexico**
>
> For over 300 years, Hispanic families in northern New Mexico have ranched, farmed, mined, and logged in the valleys and mountains in and around what is now the Cuba District of the Santa Fe National Forest. For a variety of reasons, including a lack of opportunity, few young people from this community chose careers in natural resource management. With relatively few professional staff raised in or familiar with the local culture, relationships between the U.S. Forest Service and the dominant local culture were poorly developed and often strained. Moved to address the problem by a local Hispanic rancher, who was also a Forest Service employee, the Cuba District has invested heavily to encourage local youth to consider careers in natural resource management and conservation. In partnership with the local school system, the district developed a high school program that gives students hands-on opportunities to learn about careers in natural resource management. Once a week during the spring semester, the district brings in a professional to talk about his or her field and responsibilities. In addition, field trips and encouragement to participate in the Cuba District's large Youth Conservation Corps summer jobs program inspire local youth to learn more about natural resource manage-ment careers. And, a summer camp organized by the USFS Regional Office in Albuquerque provides additional opportunities for rural teens to consider natural resource management careers. Already, the educational efforts of the Cuba District are yielding short-term benefits in the form of better community relations. More importantly, the investment should produce long-term benefits for both the agency and the community as the number of Hispanic natural resource managers grows and intercultural communication and cooperation improve.

productive way. Changing established patterns in large, bureaucratic organizations such as public land managing agencies can be difficult. Collaborative studies that allow groups to design and participate in their own research, co-management efforts that recognize the importance of traditional knowledge, and a growing acknowledgment of the importance of understanding the interaction between human cultures and natural resources are hopeful signs that this information will be put to sound use.

SELECTED READING

Beebe, J. 1995. Basic concepts and techniques of rapid appraisal. *Human Organization* 54(1): 42–51.

Ewert, A., D. Chavez, and A. Magill (eds.). 1993. *Culture, Conflict and Communication in the Wildland–Urban Interface.* Westview Press, Boulder, CO.

Harmon, D. 1987. Cultural diversity, human subsistence, and the National Park Ideal. *Environmental Ethics* 9 (Summer): 147–158.

Jones, L., et al. 1996. Howdy, neighbor: as a last resort, Westerners start talking to each other. *High Country News.* (Special issue on Environmental Collaboration), May 13.

Kemmis, D. 1990. *Community and the Politics of Place.* University of Oklahoma Press, Norman, OK.

Light, S., L. Gunderson, and C. S. Holling. 1995. The Everglades: Evolution of Management in a Turbulent Ecosystem. pp. 103–168. In: L. Gunderson, C.S. Holling, and S. Light (eds.), *Barriers and Bridges to the Renewal of Ecosystems and Institutions.* Columbia University Press, New York.

Rodriguez, S. 1987. Land, Water, and Ethnic Identity in Taos. pp. 313–403. In: C. Briggs and J. Van Ness (eds.), *Land, Water, and Culture: New Perspectives on Hispanic Land Grants.* University of New Mexico Press, Albuquerque, NM.

Snow, D., and D. Clow (eds.), 1995. The art of listening. *Northern Lights* 11(1): 9–23.

Social Classification

Why Social Classification Is Important To Ecological Stewardship

Classification of objects, persons, behaviors, events, times, places, ideas, and all other things is a ubiquitous, universal, and natural characteristic of the human species. Naming has both survival and aesthetic value. Like biological taxonomy, social classification seeks to impose order on an otherwise overwhelming spectrum of social and cultural characteristics. Naming can provide a guide as to how we should respond to that object, person, or idea. These plants are poisonous and these are edible, these are bitter and these are sweet. These persons are concerned about the environment as a source of income and these persons are primarily concerned about the environment as a setting for recreation. Classification oversimplifies reality to help us understand the complex world around us (Box 1).

This summary was drafted by Nels Johnson with contributions from Maureen McDonough, Don Callaway, Marie Magelby, and Bill Burch. It is based on the following chapter in Ecological Stewardship: A Common Reference For Ecosystem Management, Vol. III:

McDonough, M.H., D. Callaway, L.M. Magelby, and W. Burch. "Social and Cultural Classification in Ecosystem Management."

KEYWORDS: Anthropology, demography, sociology, scale

> **Box 1**
> **The Limits of Classification**
>
> The notion of a pure, value-free classification scheme in natural or social science is an aspiration rather than a reality. History provides numerous examples of social and cultural classification schemes — apartheid in South Africa, for example — that have been used to discriminate, persecute, and segregate peoples rather than to understand their similarities and differences. Even where classification is used to facilitate understanding, these schemes should not obscure the complexity of either the social or biophysical world. Nor should we believe they are reality. Classifications, of course, have a purpose or an objective. Behind each classification scheme and the theories they support are assumptions that should be clearly stated. Classification is not (or should not be) a capricious *ad-hoc* process, but one that is the basis for a systems approach to understanding social systems.

Social classification, when done well, is an important and necessary tool for ecological stewardship. Categories help us place newly encountered objects or people into groups whose characteristics we think we know and understand. As society diversifies, changes, and expands, social classification can help resource managers to understand and anticipate both problems and opportunities for sustainable resource management. Without social classification, researchers and managers may fail to adjust management and policy to better serve the needs of more people. For example, carp and other "non-game" fish species (sometimes called "trash" fish species) have been widely shunned as sports, food, or aesthetic objects by many Americans, including fisheries managers. However, carp and other "undesirable" fish species are often valued by people of Asian and other cultural origins who have immigrated to the United States (Fig. 1). Assuming that a given resource is similarly valued by all people may result in policies and practices that do not serve the interests of many. As ecosystem management challenges the way natural resource managers classify features of the biophysical environment, it should also challenge the way they classify people.

Fig. 1. Members of different ethnic groups sometimes have conflicting views on the value of a particular resource, such as carp. (Photo courtesy of U.S. Fish and Wildlife Service.)

KEY FINDINGS

Classification is the fundamental basis for scientific theory, including social science. Compared to biological taxonomies, social classification is both similar and distinctive.

✓ **A wide range of social and anthropological variables can be used in a classification system**. Biologists use different variables to construct basically two alternative taxonomies. Morphological characteristics form the basis for classical Linnean taxonomy and evolutionary pathways have been used for cladistic taxonomy. A much wider array of variables and typologies are used in social classification. A leading researcher, Kenneth Bailey, identifies 14 kinds of social classification that use either quantitative or qualitative methods (or a combination). Variables can include an enormous array of characteristics: economic, educational, and marital status; religious, political and social beliefs; ethnic identity and geographic location; kinship, court-ship, and governance. The choice of classification scheme and variables depends, of course, on the purpose for which social classification is used.

✓ **Social classification is a simplification of a continuum**. This is also true in biological classification, particularly at the community and ecosystem levels. In general, classification can become "firmer" or more precise at more local scales. The appropriate use of social classification depends on the objective and at what scale the analysis is taking place. The strength of the classification system is related to the validity of the measure. In other words, are you measuring what you think you are measuring at the appropriate scale? (Box 2).

✓ **Extrapolating a social classification system from one location to another can be misleading**. The advantage of a detailed and precise classification scheme for a well-defined and limited geographic area is that it is more likely to be predictive of human behavior. Applying that classification scheme to another geographic "community" (even neighbors), however, can be misleading or dangerous. The finer resolution of this kind of "deep" classification approach means that it will be sensitive to a range of variables

---

Box 2
Choosing the Right Scale for Social Classification

Choosing the correct level of analysis for understanding and predicting human behavior has long been recognized as a challenge in a variety of social science disciplines. The least uncertainty in predicting human action occurs at the most restricted level of analysis relative to the problem. Aggregating across multiple changes caused by multiple forest management activities in multiple locations with multiple causes creates unpredictability rather than the desired ability to predict. For example, there is some agreement that studies of natural resource dependent communities have been at inappropriate levels of analysis to account for the range of factors that influence these communities. These factors range from individual sense of place to national demands for forest products. Studies based on geographic or political boundaries, therefore, do not help us understand either the impacts of changes in resource availability (e.g. a reduction in timber supply) or community responses to these changes. In addition, information collected at the household level, for example, but aggregated to the community level runs a serious problem of specification error. These are important concerns for ecosystem management as we attempt to define the biophysical boundaries of ecosystems and incorporate human factors at an equivalent or appropriate scale.

that are likely to vary from one area to another. If the classification scheme is to be used elsewhere, it should be adapted and revised to reflect these differences. Alternatively, it may be preferable to choose a "wide" classification based on fewer variables that reasonably reflect social groupings across the broader area.

✓ **Social stratification is an important aspect of classification.** Social stratification research examines the distribution of political and economic power — who the powerful are, how they exercise that power, and the means by which they retain that power. Social power is often the most important determinant of which natural resource management strategies are implemented and which are not. The study of social power and social stratification is seldom evident in attitude surveys that ask individual opinions about likes and dislikes. Power becomes evident when a rancher who has had 400 animal unit grazing months on federal land for three generations is confronted by an agency expert who denies a continuation on the basis of biological carrying capacity. An effective and meaningful scheme of classification for ecosystem management will need to draw more upon the tradition of stratification taxonomies than simply clustering attitudes on passing issues of environmental interest.

✓ **Social classification is a dynamic process.** Societies change through cultural diffusion, and environmental and economic change. In some cases, social change is extraordinarily rapid. For example, nearly one quarter of all children in the United States in 1995 lived in households headed by single women — more than four times the percentage in 1970. Social and political beliefs can also change rapidly as newcomers move into an area, or as a younger generation of voters matures. Much more than biological taxonomies, social classification must be constantly updated to reflect changing social systems.

PRACTICAL CONSIDERATIONS

Social classification is used by both sociologists and anthropologists. There are distinctive differences between their approaches, although the major findings above apply to both. Both types of approaches can be valuable in the context of ecosystem management.

Classification in Sociology

Sociologists tend to classify people on the basis of the way they organize themselves. Human social organization can be classified both vertically and horizontally.

Vertical classifications examine social characteristics at the individual, household, and community level. Characteristics at each level can help us understand social attitudes and behavior with respect to natural resources. Classifications at only one level, say using individual characteristics to predict economic choices in isolation from community economic constraints, may limit their usefulness.

Individual classifications based on demographic characteristics are most useful as social indicators. Demographic characteristics such as age, gender, and ethnicity are relatively easy and inexpensive to acquire, and are often used in natural resource management to classify people. They can be quite useful as social indicators, but when aggregated to describe populations, they can obscure relationships between individuals that help understand why people behave the way they do. For example, studies on economic prospects in logging and mill communities in the Pacific Northwest often focus on individuals in the labor market to forecast whether people will invest in education to do something different, or migrate elsewhere to find a comparable job to the one they left behind. However, family/household and community constraints on economic opportunities may be at least as important in predicting individual choices.

Household classifications are often used to understand consumption and use of natural resources. Households are the basic unit of analysis within a community. A household is defined as a group that shares living accommodations and cooperates in meeting household needs through complementary activities, often involving the exploitation of common resources such as land. Decisions to purchase forest products are made at the household level. Studies of poverty in natural-resource-dependent communities generally focus on the differential impacts of declines in resource availability on different types of households. The number of studies on the role of forest products in household economies is increasing, including household food security studies, gathering of special forest products, and subsistence use of wildlife resources. Key to understanding natural resource use at the household level is an analysis of division of labor, including how households pool their resources both formal (e.g. employment) and informal (e.g. gathering and bartering) to develop survival strategies.

Community classification can be used to understand the potential impacts of major policy changes. The use of communities as units of analysis for examining ecosystem management issues has become widespread in recent years as agencies wrestle with the question of who will be impacted by major policy decisions. The concept of community can be defined as place based on geographic proximity, or affiliation based on common interests regardless of geography. As with households, research indicates that the ability of resource-dependent communities to cope with changes in resource availability vary within and among communities. An analysis of 300 communities affected by changes in timber availability in the Pacific Northwest identified several key characteristics related to community capacity to adapt to these changes: physical and financial infrastructure, human capital, and civic responsiveness. Civic responsiveness includes leadership (formal and informal) and institutional infrastructure, including community assistance agencies and links to institutions external to the community. Communities without these characteristics are at risk from changes in resource (e.g. timber) availability.

Horizontal classifications focus on kinship, religion, affinity, and other cultural characteristics. Within these "units" and between them, the relations among people are structured in a patterned way. Behavior is guided by rules of what is appropriate. The long standing differences between the U.S. Forest Service and Hispanic subsistence farmers in northern New Mexico over land and water rights serve as a classic example of how horizontal classification can help to understand conflicts over natural resources (see Box 3).

Classification in Anthropology

When anthropologists try to understand social behavior, they distinguish between two major perspectives. In one ("emic" classification), researchers seek to understand how people in another culture think: how do they perceive

Box 3
Cultural Beliefs, Land and Water in Northern New Mexico

For more than two centuries, significant areas of land were held communally on Spanish-American lands in what is now northern New Mexico. Each family had rights to use these lands. In contrast, land was held in private ownership by Anglo-Americans who began arriving in the late 1800s. When National Forests were established in the early 1900's, the Forest Service did not recognize traditional land uses. Spanish-American lands within forest boundaries were managed on the principle that common lands belong to no one, and common lands became national lands. Questions of land ownership were decided in Anglo courts by Anglo judges unfamiliar with different land tenure traditions. More recently, Hispanic subsistence farmers in northern New Mexico have been fighting the Forest Service over water rights. Water is an important part of cultural cohesiveness in these farming communities. Community organization is based on the management of the water ditch, or acequia. Water rights are communally owned and shared, not viewed as a commodity to be bought and sold. When an ecosystem manager who must decide on a course of action is confronted directly or indirectly by constituents bearing competing claims, classifying by shared values and norms can help distill information into a usable form. In using "cultural" categories to classify people, ecosystem managers need to be careful that groups that perceive and experience the natural environment from divergent vantage points are identified and given serious consideration. Sometimes the most important factors in predicting and understanding human behavior are cultural restrictions. Policy is not implemented when it is culturally unacceptable.

and categorize the world, and how do they conceptualize and explain things? In the other perspective ("etic" classification), indigenous categories and explanations are translated into more traditional "western" analytical categories.

Traditional Ecological Knowledge (TEK) is a useful form of emic classification. In many traditional cultures, people have accumulated detailed knowledge about their environment after centuries or millennia of living in a region. This knowledge, sometimes referred to as Traditional Ecological Knowledge (TEK), forms the basis for many practices and beliefs (Fig. 2). In Alaska, for example,

Fig. 2. Traditional knowledge systems can provide valuable information for fish and wildlife management.

traditional ecological knowledge may be as effective as, and perhaps more useful than, complex and abstract ecological models. For example, at a public meeting in northwest Alaska, western biologists had come to inform local communities of a proposed project to research the causes of a perceived moose decline. The major focus of the research design was to use airplanes to fly transects and count the moose. During this presentation, an elder quietly suggested they count beaver lodges while they are at it. This suggestion was dismissed, although later evidence clearly showed — as the elder had recognized — that an increased population of beaver (and their dams) had substantially altered the habitat to the detriment of moose. Traditional knowledge systems can have practical benefits for both research and management, and should be considered as vital sources of information where they exist.

Etic classification can be used to translate traditional views into western analytical categories. Resource managers, research scientists, courts of law, and people in their every day life use multiple classification schemes that often contain unwarranted assumptions derived from their own set of cultural values. For example, in a key ruling on a claim by Native Alaskans on damages from the Exxon Valdez oil spill in Prince William Sound, a judge asserted that their subsistence lifestyles differed from non-native Alaskans in degree but not in kind. Anthropologists using etic classification methods, however, have shown very substantial differences. For example, 388 key people in the oil spill area were asked to identify 77 naturally occurring resources (marine and land mammals, fish, plants, berries, inter-tidal invertebrates, and so forth) in the area. Researchers inquired which of the 77 species were available locally, and whether the amounts available were sufficient. Sixty-nine percent of these people were non-Native, and 31 percent were Native. Some 95% of Natives responded to all 77 questions about resource sufficiency, but not one non-native responded to all 77 questions. Despite the impacts of the spill on traditional harvesting activities, in 1991 the average household in predominantly Native communities used between 20 and 25 different species. During the same year the average household in the predominantly non-Native communities used 6 to 8 species. Clearly, despite the fact that both groups depended on harvesting wild species, the Native Alaskans used a wider variety of species that were directly impacted by the oil spill.

SELECTED READING

Bailey, K.D. 1994. *Typologies and Taxonomies: An Introduction to Classification Techniques.* Sage Publications, Thousand Oaks.

Duncan, O.D., 1964. Social Organization and the Ecosystem. In: Faris (ed.), *Handbook of Modern Sociology.* Rand McNally & Company, Chicago, IL.

Humphrey, C.R., G. Berardi, M.S. Carroll, S. Fairfax, L. Fortmann, C. Geisler, T.G. Johnson, J. Kusel, R.G. Lee, S. Macinko, N.L. Peluso, M.D. Schulman and P.C. West. 1993. Theories in the Study of Natural Resource Communities and Persistent Rural Poverty in the United States. In: Rural Sociological Society Task Force on Persistent Rural Poverty (eds.), *Persistent Poverty in Rural America.* Westview Press, Boulder, CO.

Kleindorfer, P.R., H.C. Kunreuther and P.J.H. Schoemaker. 1993. *Decision Sciences: An Integration Perspective.* NY: Cambridge University Press, New York.

Machlis, G.E., J.E. Force and J.E. McKendry. 1995. *An Atlas of Social Indicators for The Upper Columbia River Basin.* Contribution 759. Idaho Forest, Wildlife and Range Experiment Station, Moscow, ID.

Pressman, J.L. and A. Wilavsky. 1984. *Implementation: How Great Expectations in Washington are Dashed in Oakland.* University California Press, Berkeley, CA.

Social Processes

Why Social Theory Is Important To Ecological Stewardship

Resource managers have often pointed to "people" as the least understandable and predictable factor in the management of natural resources. Yet, an ecosystem approach calls for directly considering the interests of people and viewing people as part of — not apart from — their natural environment. Social science provides insight about the fundamental aspects of resource management as: (1) a social activity practiced by a specialized community of trained professionals, (2) located within specialized organizations created by humans, and (3) guided by institutions created by human law.

This summary was drafted by Andrew Malk and Kathy Parker with contributions from Victoria Sturtevant. It is based on the following chapters in Ecological Stewardship: A Common Reference for Ecosystem Management, Vol. III:

Parker, J.K., V.E. Sturtevant, M.A. Shannon, J.M. Grove, J. Ingersoll, L. Sagel, and W.R. Burch, Jr. "Some Contributions of Social Theory to Ecosystem Management."

Burch, W.R. Jr., and J.M. Grove. "Ecosystem Management: Some Social and Operational Guidelines for Practitioners."

KEYWORDS: Community, communication, conflict, social theory

People form a set of conceptual frameworks to help them understand the interaction of human and ecological systems. The construct of these theoretical frameworks depend on the needs and backgrounds of the users. Citizens can use a conceptual framework to: (1) make explicit their perception of reality; (2) express their understandings and values of ecosystems; (3) articulate their processes of interaction with each other, with other biological species, and with non-living elements of the environment; (4) provide a basis for what needs to be learned and how learning, from their perspective, takes place; and (5) provide a record of what they desire and/or anticipate as outcomes from ecosystem management interventions.

Researchers can use a conceptual framework to: (1) outline and justify assumptions they make and the questions they ask during the research process; (2) help identify the most significant variables that need to be considered, and suggest the linkages that may exist between them; (3) help guide collection of data; (4) clarify the role researchers themselves play during the course of research; (5) more explicitly link questions of citizens, managers, and policy makers in research efforts; and (6) provide a sound basis for any recommendations proposed.

Ecosystem management practitioners can use a conceptual framework to: (1) understand the realities with which they have to work; (2) understand the complex interactions between humans and their resources, and the potential impacts of given management interventions on humans; and (3) identify potential obstacles, opportunities, and options available to them as they design and implement on-the-ground, multi-scale responses for adaptive management.

Policy makers can use a conceptual framework to have: (1) a basis for raising questions and analyzing information that comes to them from researchers, practitioners, citizens, and organizations; (2) better understanding of the complex interactions and issues on which they must make decisions; and (3) more insight into the potential intended and unintended, direct and indirect impacts of policy interventions.

KEY FINDINGS

There are real benefits to managers in reaching out to the social science professional community for ideas and assistance. Several selected areas of social theory are highlighted here.

✓ **Social theory helps resource managers understand conflict**. The body of conflict theory is characterized by competing hypotheses. One set of hypotheses views conflict as destructive of trust and order. Another set views conflict as a source of creative capacity for learning within a flexible organizational framework. This difference in views, especially in explaining action and behavior, can be of great use to managers. In Nevada, the Spring Mountains National Recreation Area Collaborative Planning process demonstrated the constructive potential of conflict. In this case, conflict has been necessary for creative ideas to emerge in collaborative efforts. Although participants in the planning process fought over each word of the proposed legislation for a National Recreation Area designation, they gradually recognized a common interest in a place that was special to all of them, even if for different reasons. Parking lot shouting matches gave way to civil discussion and progress on a legislative proposal. While conflict was part of daily relationships, it was not interpersonal conflict. Differences of viewpoint, philosophy, and interest gradually grew into a vision and the means to achieve it.

✓ **Social theory helps to explain bureaucracy, administration, and technical rationality**. Federal land management agencies are required to serve as the neutral conduit for the voice of Congress, the technical analyst to achieve the broad policies given by Congress, and the conciliator among the demanding voices of citizens and interest groups locally and nationally. The Spring Mountains National Recreation Area Collaborative Planning process illustrates how people at all levels joined in formal and informal mechanisms to develop the vision of a new National Recreation Area into a reality. The Forest Service reached out to other agencies with which it often had conflicting relationships, to State and local governments, Indian tribes, and conservation groups. Land management agencies can often provide proactive leadership in building such collaborative partnerships, because they have both the tradition of informal and personal relationships within communities, and formal processes for decisions and appeals.

✓ **Social theory helps understand how organizations learn**. Theories of organizational learning focus on how organizations develop strategies for creating new organizational structures, cultures, processes, and performance criteria. Organizational learning is not just a one-time intervention. It is a transformational process with several distinct elements: commitment to becoming a learning organization; experimentation with new ways of thinking, acting, and working; systemization of effective strategies for organizational learning; and development of a new identity as a learning organization. This transformation occurs only by changing how the members of the organization think and how they interact. It is not surprising that natural resources professionals, as employees, share the cultural values and biases of their organizations. In some cases, employees are so well socialized into the cultural norms of their employing organization that their range of actions are constrained by this culture and training. Public agencies, as learning organizations, need to move from being reactive and custodial towards being action-oriented and participatory. In Baltimore, city foresters are working to understand how different types of property rights can hinder or promote urban forestry initiatives (Box 1). In the Applegate Adaptive Management Area in southwestern Oregon, agency personnel report that working with the community on specific projects (e.g. field trips and booths at community days) and in community activities oriented towards improving the watershed (e.g. tree planting and nursery plot thinning) improved communication and learning (Fig. 1). As people recognize that such efforts contribute to building social capital and solidarity, important bases for collaboration and trust, they more clearly understand why these processes are important to agency effectiveness.

✓ **Social theory helps understand communication, adoption of innovation, and technology transfer**. Although organizational theories focus on strategies for organizational transformation, the adoption/diffusion literature focuses on processes by which individuals and/or groups adopt new practices or new ideas. For example, the vast majority of studies have shown that personal friends and local leaders serve as the strongest source of influence as compared to the media. Experience with forestry certification outreach shows how important it is to put a strategy for the adoption of a new practice through a trial run. The mechanisms through which people

**Box 1
When There is No Owner**

Traditionally, foresters are taught to find the landowner, identify their goals and objectives, develop a plan, and implement it. That's it and it's pretty simple. However, I am just starting to realize how new and different working in urban areas is, because so many times in urban areas there may not be a landowner. In this case, we have to figure out what the community's goals and objectives are and then who the owner or owners should be. It's a complete reversal of how we have been taught to work and to do business and we need to learn how to do it.

Gene Piotrowski
Chief of Urban Forestry
Maryland Department of
Natural Resources

Fig. 1. A key to developing partnerships is getting individuals into the field to participate in the design of projects. The Applegate Partnership in Oregon does this regularly. (Photo courtesy of Sue Rolle.)

are contacted can impact their responsiveness and willingness to even find out more about potentially useful technologies. Incorporation of local input has been identified around the world as a critical factor in the successful adoption and application of technology. Local people often have better empirical knowledge about the object or process of study than a manager; this knowledge represents a learning opportunity for the manager.

✓ **Finally, social theory helps to understand different kinds of communities**. This area of theory illuminates a range of issues central to involving the public in ecosystem management. For example, we know that there are many kinds of communities. The community of occupation crosses certain boundaries through the link of common work and common life expectations. Communities of expertise link managers to the academic and scientific enterprises responsible for creating access to community resource information. The community of place remains a strong orienting framework for many people living in both rural and urban settings, as the landscape serves as an important spatial basis of rural identity, social relationships, and public life. Communities of interest are geographically dispersed, but socially well-integrated, regional and national groups. Most political associations are based more on the community of interest. Tension between communities of place and communities of interest is an important quality of local community life. In any setting, one can expect to find all of these community types. The Applegate Partnership holds lessons for public agencies whose goals will be achieved in part by melding communities of interest into a single community of place. By assembling partnership groups, whose members agreed to put their interests aside in the interest of

Fig. 2. The Applegate Partnership meets regularly to discuss a wide range of issues. Meetings are open to all who are interested and are designed to encourage the exchange of information between participants. (Photo courtesy of Sue Rolle.)

the collective and ecosystem, agencies helped communities of interest move to being communities of place (Fig. 2). What starts to bring members together is a shared attachment to a common place.

PRACTICAL CONSIDERATIONS

Social scientists have looked at natural resource management issues for a long time. There are a wide array of social theoretical "lenses", that address social and organizational contexts where individuals and organizations come into conflict, where bureaucrats interact with various publics, where community members of different backgrounds address issues of mutual interest from a variety of perspectives, where public organizations are being asked more and more to become learning organizations with strategies that provide them with more adaptability and concomitant accountability, and where communication strategies must be more targeted to the audiences they intended to reach. Managers can have a more relevant angle of vision for analyzing people and ecosystem interactions.

An array of social science methods, tools and techniques (e.g., rapid assessment of tenure or property rights) can be used for action research. Knowledge obtained from these tools contribute to increased understanding of human institutions and how they may or may not affect ecosystem management activities.

The dynamics of societies and groups vary with context, scale, and over time. External and internal forces (e.g. national and local politics) affect relationships (who is a stakeholder and how does s/he relate to other stakeholders), roles (e.g., who has legal authority, who has traditional authority), dynamics of interactions (competition, cooperation, collaboration), and the direction and magnitude of change. Temporal aspects of a social context also can help managers. For example, understanding different stages of the social contexts in which they are working can provide managers with insights into the kinds of interactions they might need to encourage. At an earlier stage more time and energy may be required for building trust and identifying and obtaining input from a broader set of stakeholders. At a later stage, efforts may be more directed at maintaining existing relationships and working with elected or designated representatives. Society and social groups are dynamic entities. The better managers understand and respond to the dynamics over time, the more effective they will be.

Semantic issues are the foundation of many conflicts that might otherwise be minimized if everyone had a common frame of reference if not a common frame of agreement. Definitions serve as a starting point for discussion of concepts that may be viewed in different ways by different people. Scientific efforts, especially of an interdisciplinary nature must address semantic issues that are reflected in basic concepts: determination of what should be considered a "problem", the basis of categorizing, commonly held images and metaphors, and what makes one set of arguments more persuasive or decisive than another set, for example. In other settings, the semantics and social content of even something biophysical, like a forest, must be understood from any of a number of perspectives. A tourist might think of the forest and its immense trees in terms of their grandeur, while employees working in a local lumbermill might think of them in terms of the number of years of work and income they will provide their families. Fundamental differences in the way humans think about, talk about, and interact with the biophysical world and with each other, therefore, must be better understood as a starting point for continuing dialogue and decision making.

"Social capital" provides a foundation for understanding social relationships. "Social capital" is not traditional capital (money and what it buys), nor is it human capital (know how and skills); or assets (infrastructure). Social capital includes the resources of human relationships that usefully connect people in more trusting group and organizational relationships. Building social capital in an organization, an association, a town, or other social group means developing a strong, viable, and sustainable source of consciousness, creativity, and community. By focusing efforts on learning how to build and encourage social capital, perhaps ecosystem managers can better understand one of the primary aspects of sustainable development: the ability a community to deal with public disputes, opportunities for economic and social growth, allocation of scarce resources, and other potentially contentious decisions that have to made. Successes with various kinds of cooperation, coordination, and collaboration can serve as templates (not blueprints) for future interactions.

SELECTED READING

Berry, J.K. and J.C. Gordon (eds.). 1992. *Environmental Leadership:Developing Effective Skills and Styles.* Island Press, Washington, DC.

Burch, W.R. Jr. 1992. Our theme: thinking social scientifically about agroforestry. In: W.R. Burch Jr. and J. Kathy Parker (eds.), *Social Science Applications in Asian Agroforestry. Winrock International, USA and South Asia Books, USA, New Delhi.*

Cernea, M. (ed.), 1985. *Putting People First.* Oxford University Press, Washington, DC.

Field, D.R. and W.R. Burch, Jr. 1988. *Rural Sociology and the Environment.* Greenwood Press, New York.

Lee, R.G., et al. (eds.). 1989. *Community and Forestry: Continuities in the Sociology of Natural Resources.* Westview Press, Boulder, CO.

Miller, S.E., et al. 1994. *Rural Resource Management: Problem-Solving for the Long Term.* Iowa State University, Ames, IA.

Economic Dimensions

♦ *Shifting Human Use and Demands for Natural Resources*

♦ *Economic Interactions at Local, Regional, National and International Scales*

♦ *Ecological and Resource Economics*

♦ *Uncertainty and Risk Assessment*

♦ *Economic Tools for Ecological Stewardship*

Demographics and Shifting Land and Resource Use

Why Demographics and Shifting Land and Resource Use Are Important To Ecological Stewardship

By the numbers, humans are perhaps the most successful vertebrate species in the history of life on earth. Human numbers grew slowly at first. It wasn't until 1830 that the population reached one billion — approximately one to two million years after *Homo sapiens* evolved as a distinct species. By the 19th century, however, the human population was rapidly expanding. It took only a century for the population to add another billion, and then this figure doubled to four billion between 1930 and 1974. The most recent billion was added in the last 12 years. Today, the world's population increase in one year (90 million) is roughly equivalent to the total population growth between 1600 and 1700. This rapid relative and absolute increase in population stretches the productive, absorptive and recuperative capacities of the Earth. It also stretches human capacities for technological and social invention, adaptation and compassion.

This summary was drafted by Nels Johnson with contributions from Steven Cinnamon. It is based on the following chapters in Ecological Stewardship: A Common Reference for Ecosystem Management:

Cohen, J.E. "Human Population Growth and Tradeoffs in Land Use" (Volume II).

Cinnamon, S.K., N.C. Johnson, G. Super, J. Nelson, and D. Loomis. "Shifting Human Use and Expected Demands on Natural Resources" (Volume III).

KEYWORDS: Population, economics, culture, resource consumption, technology, information, media

Understanding the relationships between population growth and land use are important for several reasons. Planners, managers, and citizens must consider the global perspective, even if they are concerned only to protect American resources and interests, because the United States is intimately linked to the rest of the world. The United States is linked demographically to populations abroad through migration and competition for jobs; economically through international markets and international technologies that affect the demand for commodities and services derived from land; environmentally through atmospheric emissions, introduced weed and pests, and global climatic changes; and culturally through the spread of free-market institutions, rising material expectations and consumerism, technologies, political movements, and other values that affect the supply of and demand for products and services derived from land and water.

KEY FINDINGS

Population is an important determinant of land and natural resource use, but it is not the only factor. Understanding the relationship between population and other factors is key to understanding the dynamics of resource and land use changes.

✓ **Human population growth interacts with economics, the environment, and culture to determine resource use**. A *population* is described by its size (the numbers of people by categories of age, gender and other characteristics), rate of growth or decline, spatial distribution (for example, urban versus rural) and migration. *Economics* includes institutions for ownership or common use of land, incentives for land exploitation or conservation, markets or other institutions for dealing in land as well as the products of and inputs to land, labor force availability, and sources and conditions of capital and credit. The *environment* includes the physical, chemical and biological quality of land, air and water, including climate. *Culture* includes political institutions; governmental, commercial, and individual policies toward land use; styles of life; expected roles of women, men, children and elderly in paid work and family life; levels of education; and religious and traditional views of relations between humans and their land and water. The relationship between population and resource consumption and land use is mediated by these other factors (Box 1). The bottom line here is that population alone does not determine land use patterns. Rather population in combination with economic, environmental, and cultural factors helps to determine resource consumption and ultimately land use.

✓ **The rate of global human population growth peaked between 1965 and 1970, but the total population will continue to grow for decades**. Human populations have grown steadily in both absolute and percentage terms throughout history (although massive epidemics in the 14th century, and possibly earlier, briefly interrupted the increase). Between 1965 and 1970, the annual growth *rate* peaked at 2.1 percent — a doubling time of only 33 years! Still, even with a decline in growth rates, every year between 1970 and 1990 saw more numbers added to the population than had arrived the year before. The absolute growth in population did not peak until the early 1990s when it crested at approximately 93 million new faces every year. The United Nations projects the total population will continue growing from the

---

Box 1
Population Alone is an Unreliable Predictor of Resource Use.

During the course of a year, a person living in Shanghai, China uses about 50 pounds of paper. This includes not only stationary, newspapers, and tissues, but packaging, boxes, and food containers. Meanwhile, someone living in Denver, Colorado will go through approximately 700 pounds of paper in a year. China, with more than one fifth of the world's population, consumes about seven percent of the world's paper production. The United States, with less than 5 percent of the world's population consumes more than a quarter of its paper production. Why the big difference? Clearly, population size alone cannot explain the difference. The answer lies in some combination of economic, environmental, and cultural factors. For example, with a per capita income only a fraction of that in the United States, most Chinese simply can't afford to purchase as much writing paper or buy as many goods packaged in cardboard boxes as the average American. With most of its limited fertile lands devoted to food production, China's environment is simply unable to produce the volume of wood fiber that is produced in the United States. Cultural attitudes also shape different views toward the use of paper by people living in China and the United States. However, as China's economy grows faster than its population and as its citizens become more culturally exposed to Western-style consumerism, its total paper consumption may rival that of the United States within the next decade.

---

current 5.8 billion until it reaches a peak of approximately 8–9 billion sometime during the middle of the next century (Fig. 1).

✓ **Population growth is unevenly distributed**. Global statistics conceal very different stories in different parts of the world. About 1.2 billion people live in the economically more developed regions, where average annual incomes are $18,100: Europe, the United States, Canada, Australia, New Zealand, and Japan. The population of the rich countries increases very slowly (about 0.1% per year) — a doubling rate of 500 years. The remaining 4.6 billion live in the economically less developed regions, where average annual incomes are $1,100. At current growth rates (1.9%), the population of developing countries will double in 37 years. The growth rate is especially

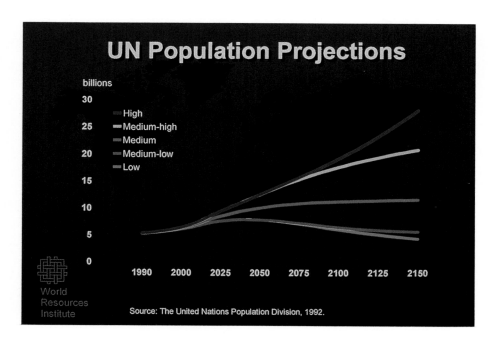

Fig. 1. Population projections by the United Nations vary widely. Mid-range projections indicate a peak population of approximately 9–10 billion in 2050.

Fig. 2. Population growth rates are highest in poor developing countries. In Madagascar, pictured here, the current population of 16 million is projected to double in less than twenty years (Photo credit: Kirk Talbott).

high in Africa (2.8%), which is also the world's poorest region (Fig. 2). In aggregate, the richest one-fifth of the world's population generates and spends about 80% of the world's income. One notable exception to this pattern is the United States. The annual growth rate in the United States averages just under one percent. While this is a modest growth rate compared to many developing countries, it is among the highest in developed countries. Over the next 3 decades, the U.S. Bureau of the Census projects the population to continue growing at an average of about one percent (mostly as a result of immigration) — or an increase of 79 million — to 346 million in 2030. Growth rates within countries vary as well (Box 2).

Box 2
Where Are They Going?

There is substantial regional variance in population growth within the United States, mostly due to internal migration from one part of the country to the other but also due to the settlement patterns of immigrants from other countries. For example, while the population in northeastern states grew by just 1.4 percent between 1990 and 1997, the Rocky Mountain states experienced population gains of 20.2 percent. Locally, some parts of the West are experiencing nearly exponential growth —since 1990 Las Vegas, Nevada has grown from 350,000 residents to 1.2 million! Meanwhile the Pacific coast, south central and southeastern states grew at more than twice the rate of midwestern and northeastern states. These regional differences in population growth continue trends that have been in evidence for decades. That is, the U.S. population is steadily moving south and west. In 1800, half of the population lived north and east of Kent County, Maryland. By 1900, the mean center of population stood nearly 700 miles west and south at Columbus, Indiana. Since 1900, the mean center of population has shifted another 340 miles southwest to Pulaski County, Missouri. Today, nine of the ten fastest growing states are west of the Mississippi. Three out of four Americans live in urban areas, and half of all residents live within 80 km of the east or west coast.

Box 3. Technological Innovation and Natural Resource Use

Technology plays an important role in the equation linking population and demographic patterns to natural resource use. Some technological innovations may lead to greater resource consumption, others to less consumption, while yet others merely switch the type of resources that are consumed. For example, contrary to expectations, the wide availability of personal computers and printers increased demand for paper during the 1980s and 1990s. On the other hand, new recycling technologies slowed demands for "virgin" tree fiber in the manufacture of paper during this same period. And, advances in plastic polymers led many packaging manufacturers to switch from paper, glass, or aluminum to oil-based materials. How these technological innovations will affect natural resource use is one of the wildcards in predicting future human demands for natural resources and their impacts on the environment.

✓ **Demographic and economic trends will pose Americans with future tradeoffs in land use.** One reason is that domestic trends — a growing population, increased economic demand for timber and other resource commodities, and rising social demand for conservation and recreation — will likely sharpen rather than diminish conflicts between competing interests. Another reason is that the United States is becoming progressively less isolated from the rest of the world. As populations and economies grow in resource-poor countries, people around the world are likely to become larger contenders with Americans for the products and services of ecosystems in the United States (just as Americans are major consumers of natural resources elsewhere). While past expectations about the future have usually been wrong, it seems prudent to assume that more people combined with economic growth will result in more conflicts and tradeoffs in resource consumption and land use (how much, which kind, and where?).

✓ **Changes in cultural values, technology, and policy can change the equation linking demographic and economic growth to natural resource use.** The future of natural resources management will depend as much on culture and politics as it will on science, population and economics. For example, if a majority of consumers are willing to pay higher prices for products that cost more to produce but generate fewer unwanted ecological impacts (or to reduce "externalities" as economists put it), consumption levels and environmental impacts might decrease even as population and economic growth continue. And, technology can have significant, but hard to predict, impacts on land and resource use (Box 3).

PRACTICAL CONSIDERATIONS

Global demographic, market, political, economic, and environmental trends (such as climate change) are outside a field manager's control. There is no one "right" way to anticipate the local impacts of global trends. Still, natural resource managers can use a few basic strategies to help anticipate whether, where, and when national and international trends and events might affect local resources.

The news media are a convenient, if erratic, source of information on contemporary natural resource and environmental issues. The local, regional, and national media generally cover such issues more frequently than a decade or two ago. While coverage is often superficial, it does provide a rough barometer of issues that concern the public. Unlike academic reports and quantitative data, media articles or broadcasts can provide a clear human and policy context for contemporary natural resource management issues. Many important trends and issues, however, are not covered by the media or are covered only after they have created a serious problem.

Colleagues, neighbors, and local residents can be an invaluable source of information on relevant international, national, and regional trends. Published information, no matter how well researched and peer reviewed, cannot replace local knowledge. Visiting with neighbors, talking to visitors, and seeking out local "authorities" on natural resources and social and economic trends are invaluable sources of information. This local knowledge can help to confirm which national and international trends are likely to be important locally, and which trends could have the most impact. Local knowledge can also be useful to identify trends that require additional research in order to better understand potential local impacts (see below). One effective strategy for engaging colleagues and neighbors is to use widely seen media accounts of population, economic, or environmental trends to stimulate discussion.

There is an enormous, rapidly growing, and increasingly accessible array of information on demographic, economic, and environmental trends. These include regularly updated publications, journals, and the Internet. Useful and readily available reports include the *CEQ State of the Environment Report, U.S. Forest Service RPA Report* (and associated background reports), *U.S. Census Bureau Annual Reports, State of the World* (Worldwatch Institute), *World Resources Report* (World Resources Institute), and the *UN Human Development Report.* A variety of journals are also good sources of information and analysis. While access to information has been and still is a problem for resource managers living in rural areas and small communities, the Internet is rapidly breaking down many access barriers. If you have access to the Internet, chances are you can readily find an array of information on demographic, land-use, economic, and environmental trends.

Questions raised by media accounts, discussions with colleagues, and a review of readily available data may lead to research topics. If you work for an agency with research capacity, it may be useful to consult with research staff to determine whether they can project trends and assess possible local impacts. Ideally, they should work closely with resource managers to help identify possible management responses. In some cases, a relatively brief analysis from existing data (from sources such as those described above), may be useful. In other cases, it may be necessary to design site-specific surveys to collect primary data that can lead to the identification of management options.

SELECTED READING

CEQ. 1997. Council on Environmental Quality State of the Environment Report 1994–95. President's Council on Environmental Quality, Washington, DC.

Cohen, J.E. 1995. *How Many People Can the Earth Support?* W.W. Norton, New York.

Daily, G.C. (ed.). 1997. *Nature's Services: Societal Dependence on Natural Ecosystems.* Island Press, Washington, DC, and Covelo, CA.

Diamond, H.L., and P. Noonan (eds.). 1996. *Land Use in America.* Island Press, Washington, DC.

Easterbrook, G. 1995. *A Moment on Earth. The Coming Age of Environmental Optimism.* Viking Penguin Books, New York.

Laidlaw, R.M. 1993. Strategic planning: agency responses to changing management environments. In: A.W. Ewert, D.J. Chavez, and A.W. Magill (eds.), *Culture, Conflict, and Communication in the Wildland–Urban Interface.* Westview Press, Denver.

Pfister, R. E. 1995. Ethnic Identity: A New Avenue for Understanding Leisure and Recreation Preferences. In: A.W. Ewert, D.J. Chavez, and A.W. Magill (eds.), *Culture, Conflict, and Communication in the Wildland–Urban Interface.*

Ryan, J.C. and A. Durning. 1997. *Stuff: The Secret Lives of Everyday Things.* Northwest Environment Watch, Seattle.

Turner, B.L., W.C. Clark, R.W. Kates, J.F. Richards, J.T. Mathews, and W.B. Meyer (eds.). 1990. *The Earth as Transformed by Human Action: Global and Regional Changes in the Biosphere Over the Past 300 Years.* Cambridge University Press with Clark University, Cambridge, UK.

U.S. Bureau of the Census. 1998. Statistical Abstract of the United States 1998. U.S. Government Printing Office, Washington, DC.

USDA. 1995. RPA Assessment Update of the Forest and Rangeland Situation in the United States — 1993. U.S. Department of Agriculture Forest Service, Washington, DC.

Worldwatch. 1998. *State of the World 1998.* W.W. Norton & Company, New York.

WRI. 1998. *World Resources 1998–99: A Guide to the Global Environment.* World Resources Institute, U.N. Environment Programme, UN Development Program, and the World Bank. Oxford University Press, New York.

Economic Interactions at Local, Regional, and National Scales

Why Economic Interactions Are Important To Ecological Stewardship

Decisions about natural resource management inherently involve tradeoffs, often at different scales. For example, a decision to log or mine a watershed may decrease water quality for urban residents downstream, while protection of the watershed may increase pressures to extract resources elsewhere. Economic analysis is a powerful tool that can help identify stakeholders near and far, evaluate tradeoffs between alternative management goals, and indicate least-cost approaches for achieving ecosystem management goals.

This summary was drafted by Robert Szaro with contributions from Amy Horne, Roger Sedjo, and Nels Johnson. It is based on the following chapters in Ecological Stewardship: A Common Reference For Ecosystem Management, Vol. III:

Sedjo, R.A., D.E. Toweill, and J.E. Wagner. "Economic Interactions at Local, Regional, and International Scales."

Horne, A.L., G. Peterson, K. Skog, and F. Stewart. "Understanding Economic Interactions at Local, Regional, National and International Scales."

KEYWORDS: Commodities, economic analysis, tradeoffs, values

Box 1
Economics is About Value, But Not Necessarily About Money

Economics is a discipline that seeks to understand how individuals and societies allocate scarce resources to meet their needs. Resource and environmental economics is a subdiscipline that focuses on natural resource availability and use. Some natural resources, such as timber or silver, are traded in commodities markets and are therefore relatively easy to analyze under different management regimes. Markets, however, aren't perfect. Externalities, such as air pollution from silver smelting, usually aren't included in the market price. Moreover, many ecosystem services, such as the maintenance of water quality or biodiversity, are not traded at all in the traditional marketplace. Still, these resources are often highly valued by society. Economic tools can be used to help measure the value of non-marketed natural resources. For example, travel-cost methodologies measure how far people will go and how much they will pay to experience wilderness or to view wildlife. While these tools are imperfect, they can be used to help policy makers, the public, and resource managers integrate both market and non-marketed resources into policy decisions and investment options.

Understanding economic interactions in a broad context is important to ecological stewardship for several reasons. One reason is that there is a limited supply of most natural resources — whether they are marketed, such as timber, or not, such as a rare fish species. Economics can be used to assign value to very different resources and help resource managers and the public to allocate these scarce resources (Box 1). Another reason is that the economic consequences of natural resource decisions can be felt in places far removed from where the management action takes place, just as ecological consequences can. A third reason is that knowledge about likely economic interactions can help to identify political solutions to conflicts over resource management. Fourth, a better understanding of economic interactions can help resource managers identify realistic "outputs" in the context of the broader economy and public values. Finally, an understanding of economic interactions can help policy makers and resource managers use budgets and personnel more efficiently to provide the natural resources and environmental services that society values most.

KEY FINDINGS

The following points about economic interactions are important for resource managers and policy makers to consider.

✓ **Economic systems, like ecological systems, are often linked over large areas and at local, regional, national, and international scales** (Box 2). Events in one place can have implications near and far. For example, the recent economic crisis in Asian wood importing countries, such as Japan, Thailand, and Korea, has virtually shut down timber exports from some timber exporting countries such as Indonesia, the Solomon Islands, and Papua New Guinea.

✓ **For some ecosystem outputs, price effectively communicates changes in supply and demand**. Ecosystem management produces an array of "outputs," a few of which are actively traded in markets, such as timber and livestock forage (Box 3). Changes in supply in one region can affect local prices for these commodities, and create price differences between local and

Box 2
Economics and Ecology

There are many similarities between economics and ecology. Both disciplines use a "systems approach" in conceptualizing problems, emphasizing the interconnected nature of disturbances to the system under consideration. In both economics and ecology, local disturbances may have (often unpredicted) impacts on associated systems that range far beyond the intended target. Thus the ecologist must carefully determine the appropriate scale of inquiry to ensure all relevant effects are considered. Is the relevant system a small microcosm, perhaps involving local fungi, or is the appropriate scale a tropical ecosystem the size of the Amazon?

To the economist, at the broadest level all economic activity is related, directly and indirectly, to all other economic activity. This systems approach is known as "general equilibrium" economics. By this view, changes in the local economy driven by forcing factors (e.g., an increase in local demand or a decrease in authorized local timber harvests) can have subsequent effects far beyond the local economy. Just as ecologists must decide at what scale to undertake their investigations, so too must economists. Is the relevant scale the local lumber market or the world industrial wood market? To what extent does an investigation of the forest sector need to concern itself with details of wood using sectors such as the construction and paper industries?

international markets. This difference sends a signal to producers and consumers who are respectively seeking to minimize their costs of production and consumption. Higher regional prices may decrease regional production as producers and consumers exploit lower regional prices elsewhere. Decreased timber supplies in the Pacific Northwest during the early 1990s created higher prices in that region. This led to the substitution of Pacific Northwest timber by non-timber alternatives (e.g., steel studs) and increased timber imports from Canada and other countries. Among the tradeoffs to be considered in such a situation are the environmental impacts of the substitutes — e.g, increased energy use for steel, or intensive harvesting in the southern U.S., Chile, and Russia. The question thus may not be "to harvest or not to harvest?" but rather "where and what to harvest?".

✓ **Most ecosystem outputs are not traded in markets, but economics can provide useful information for policy and management decisions**. The vast majority of ecosystem outputs are not competitively priced in markets (Box 3). This makes it more difficult for managers to integrate these resources into management decisions, especially when economics is an important factor. While ecosystem services and conditions operate outside of the traditional price system, they provide very important benefits. A complete analysis should consider their specific contributions. There are various techniques for assessing these contributions (see Ecological Economics Summary). Public policy makers and resource managers can incorporate these values into their actions in a variety of ways — from allocating their budgets and personnel towards highly valued non-marketed resources, experimenting with the creation of new markets (e.g., charging movie producers for the use of national park locations), and using policy tools (e.g., tax policy) and economic incentives (e.g., subsidies) to protect non-marketed resources.

✓ **Examining only jobs and income inadequately addresses questions of trade-offs**. Job and income analysis is important for considering economic impacts related to natural resource management decisions, especially for local economies. However, many costs and benefits do not directly involve local jobs and incomes. For example, actions to intensify commodity

Box 3
Outputs from Ecosystem Management

- Biological diversity
- Carbon storage
- Desired fire frequency and severity
- Desired land proportion in various vegetation groups
- Existence of threatened or endangered species
- Healthy aquatic and terrestrial ecosystems
- Livestock forage
- Old growth forests
- Recreation
- Road access
- Scenic quality
- Soil productivity
- Special forest products
- Timber
- Unroaded areas
- Visibility
- Water quality
- Watershed connectivity
- Well-connected mosaic of high quality habitats

production in a sensitive watershed should weigh the benefits of increased jobs against the costs of mitigating water quality damages by users downstream. Local residents also value a wide range of non-income benefits from their environment that will not be captured in a simple jobs and income analysis. Finally, since local employment and income analyses tend to focus on near-term impacts, they often neglect a variety of potential economic impacts that may be felt by future generations.

✓ **Many economic and ecological costs and benefits have impacts beyond regional economies and ecosystems.** Market and nonmarket goods and services from a given supply area can provide a variety of goods and services to a host of consumers in different geographic areas. For example, the existence of roadless areas in the West provides benefits to people in the East, some of which may accrue locally as visitors travel to the area. And, since global markets exist for most commodities, policy and market signals in distant places can have substantial local effects. For example, a policy decision to use low-sulfur coal in an Asian country may generate jobs (and environmental damages) in Wyoming. Since most ecosystems provide a variety of economic products and environmental services, a given management decision is likely to initiate multiple economic interactions at a variety of scales. Because markets for natural resources are increasingly globalized, lowering production in one region to reduce environmental damages can simply transfer those damages elsewhere (Fig. 1). In the case of timber, the effect of increased regional prices caused by harvest restrictions can "ripple" through the global trading system, shifting harvest patterns to regions that may be more environmentally sensitive. In this case, the environmental benefits accruing to a particular region (e.g., the Pacific Northwest) may not be net global benefits if the costs incurred elsewhere (e.g., Russian Far East) are greater. In the case of recreation, the closing or loss of a blue ribbon trout fishing stream may affect sportsmen thousands of miles away, increase fishing pressures on remaining streams, and drive up the cost and lower the quality of the recreational activity. By understanding economic interactions at various scales associated with management actions taken in one place, managers are better equipped to build consensus around acceptable trade-offs.

Fig. 1. Regional shortages of natural resources such as timber can lead to increased and possibly unsustainable harvest somewhere else. (Photo courtesy of U.S.D.A. Forest Service.)

✓ **There is a gap between the theoretical promise of economic methods and their practical application in ecosystem management.** New economic methods and tools are helping to identify tradeoffs in ecosystem management decisions made in both the public and private sectors. For example, reflecting environmental degradation and resource depletion in national income accounts (i.e., "green accounting") is gaining acceptance as an important tool to measure a nation's wealth and its ability to provide future environmental, social, and economic benefits. At a more local scale, full-cost accounting methods can more clearly identify the range of costs and benefits associated with a management decision. In the field, conservation groups, government agencies, and the private sector are devising innovative economic incentives to help meet sustainable resource management and conservation goals. Still, a number of obstacles limit the use of economic methods in ecosystem management. These include:

- Lack of markets for most ecosystem products and services, even those with substantial economic and social benefits;

- Lack of quantitative information about many ecosystem outputs;

- Lack of standardized and widely accepted methodologies to value non-market resources;

- Lack of economic incentives in the private sector to provide non-marketed natural resources and ecosystem services;

- Lack of clearly defined ecosystem management goals and desired ecosystem outputs in the public sector, and;

- Intrinsic difficulty measuring or valuing some ecosystem management objectives such as reintroducing "natural" fire regimes or restoring habitat connectivity.

PRACTICAL CONSIDERATIONS

Natural resource managers can use a basic understanding of economic interactions to address some very practical questions in ecosystem management projects. These include:

Which Stakeholders Will Benefit Or Lose? Understanding the economic characteristics of an ecosystem output helps identify the spatial distribution of those who stand to gain or lose when more or less of that output is produced. For example, economists working on the federal interagency Interior Columbia Basin Ecosystem Assessment gathered information on who benefits from the existence or production of several ecosystem outputs. Even though only four outputs were measured (cattle grazed, road densities, timber harvest, and acres of roadless areas), the results showed that the benefits were widely distributed both inside the Basin and throughout the United States, to rich and poor, blue collar and professional. Most individuals benefited from more than one output. In other words, tradeoffs in management choices are rarely as simple as interest groups on either side of a controversial issue portray.

What are the Explicit Tradeoffs in Management Choices? Economic valuation of both marketed and non-marketed outputs from an ecosystem enables resource managers to have a better idea of the tradeoffs involved in management choices. Estimates of the value of 15 different ecosystem outputs — including timber and range and ten types of recreation — in the Columbia

Basin Assessment allow planners to see the relative economic importance of those outputs. Across the entire Basin, recreation far outweighs timber and range values. But, looking at these values at different scales is critical to management choices. For example, timber and range values substantially exceed all recreational values except fishing in the Columbia Platau region. In the North Cascades region, hiking alone has greater economic value than timber, and other recreational values such as camping and day use in developed areas are growing in importance. In the Columbia Plateau region, managers may want to emphasize assessing possible tradeoffs between timber and fishing. In the North Cascades, tradeoffs between types of recreational activities may be more relevant.

Which National And International Trends Might Affect Regional Conditions? The Northern Forest Lands Council was created in the 1980s to assess actions that could be taken to "maintain traditional patterns of land ownership and use" in northern New England and upstate New York. An important part of the vision generated through meetings with thousands of residents was a desire to see both a vibrant timber economy and preservation of open space for wildlife and recreation. Economic analyses indicated that national and international factors were likely to be more important than regional factors in maintaining a healthy forest industry and open space. For example, global competition in pulp and paper markets means companies in the region face pressures to lower production costs through greater economies of scale and sell off less productive lands. Meanwhile, demand from residents outside the region for vacation homes and recreational development was projected to grow rapidly. These two trends intersect when timberland owners find valuable recreational markets for less productive lands. Analyses of national and international trends have proven critical to informing policy makers about which options — targeting voluntary purchases of less productive timberlands for open space and wildlife conservation, for example — are likely to be most effective in achieving goals defined in the Northern Forest Lands Study.

What Are the Least-Cost Options for Reaching Ecosystem Management Goals? Understanding economic interactions provides an opportunity to design new approaches that achieve multiple goals more efficiently. One way to meet ecosystem management goals on public rangelands is to use economic signals to improve rangeland and riparian conditions. For example, instead of charging fees on a flat basis per Animal Unit Month (AUM), the U.S. Forest Service could charge fees inversely related to rangeland and riparian conditions: the poorer the rangeland condition the higher the fee. The rancher has an economic incentive to improve conditions on his own without agency intervention. The costs of such an approach, even including monitoring and adjusting fee schedules, could be less than alternatives such as retiring leases and using public resources to restore degraded range and riparian areas. Not only could such a strategy represent a least-cost option for achieving desired management goals on public lands, but it could also improve relationships between public land managers and private users of those lands.

SELECTED READING

Bowes, M.D. and J.V. Krutilla. 1989. *Multiple-Use Management: The Economics of Public Forestlands.* Resources for the Future, Washington, DC.

Dixon, J.A. and P.B. Sherman. 1989. *Economics of Protected Areas.* Island Press, East-West Center, New York, NY.

Laarman, J.G. and R.A. Sedjo. 1992. *Global Forests: Issues for Six Million People.* McGraw-Hill Inc., New York, NY.

Frederick, K.D. and R.A. Sedjo (eds.), 1992. *America's Renewable Resources.* Resources for the Future, Washington, DC.

Oates, W.E. (ed.), 1998. *The RFF Reader in Environmental and Resource Management.* Resources for the Future, Washington, DC.

Sedjo, R.A., A. Goetzl and S.O. Moffat. 1998. *Sustainability of Temperate Forests.* Resources for the Future, Washington, DC.

Ecological and Resource Economics

Why Ecological And Resource Economics Are Important To Ecological Stewardship

Ecology and economics share their origins as words in the Greek term, *oikos*, or house. Traditionally, except for their etymology, ecology and economics have not shared a great deal in common. Ecologists studied the relationships between living organisms and their environment. Economists studied the production, distribution, and consumption of human wealth. Ecological economics seeks to overcome this gap by focusing on the relationships between human economies and ecosystems, exploring their interdependence. Ecological economics emphasizes that human decisions affect ecosystem health, which has long-term significance for economic health. It recognizes that ecosystems, and the ways economies depend on ecosystem services, are complex. This suggests that any ecosystem management issue could potentially involve many stakeholders whose interests should be included in the decision process. Since ecosystem change can profoundly affect the ecosystem services that economies depend on, preserving healthy and resilient ecosystems is important for long-term sustainable economic development. If maintaining natural capital is important to ecosystem services and to economic health, those wishing to alter natural systems upon which human well being depends should shoulder the burden of proof before such alterations begin.

This summary was drafted by Robert Szaro with contributions from Stephen Farber and David Jaynes. It is based on the following chapter in Ecological Stewardship: A Common Reference for Ecosystem Management, Vol. III:

Farber, S. and D. Bradley. "Ecological and Resource Economics as Ecosystem Management Tools."

KEYWORDS: Ecological economics, ecosystem services, natural capital, valuation, costs, benefits

The concepts developed in ecological economics are important for ecological stewardship for several reasons. First, the more expansive definition of economic value recognized in ecological economics makes it easier to consider a range of environmental and social values that traditionally have been ignored by conventional (or neo-classical) economics. This enables the public and natural resource managers to better understand the trade-offs between management choices. Second, ecological economics recognizes biophysical limits to what the environment can produce (or absorb by way of disturbance or pollution) while conventional economics tends to focus on limits of financial and technological capital and labor. Recognizing the limits of natural capital provides a more realistic framework for economic decisions that affect the environment. Third, ecological economics takes a longer term view of the costs and benefits of decisions than conventional economics, which tends to discount the value of future benefits and costs. This longer term view is prompted by the recognition that there can be no technological or capital substitutions for many ecosystem services, and by a belief that actions today should not foreclose choices for future generations because we have no way of anticipating their preferences.

KEY FINDINGS

| Box 1 |
Economics In a World of Growing Demands

Economic pressures on ecosystems will only intensify in the future. Growing population levels, settlement patterns, and increased incomes will raise the demands for ecosystem resources and their services. The pressure to transform ecosystem natural assets into marketable commodities, whether by harvesting and mining resources or altering landscapes through development, is enormous and will likely grow in coming decades. Ecosystem management can establish means for assuring that these natural assets are used in a manner that provides high returns to human welfare, and sustains their abilities to continue generating valuable product and service flows for future generations.

✓ **Ecological economics has emerged as an important variant of conventional economics**. Ecological economics expands the focus of economics to incorporate biophysical realities and complex ecosystem processes into economic decisions (Box 1).

✓ **The long run sustainability of economic systems requires sustainability of the underlying natural capital on which those economies rely**. Maintaining the health of ecosystems, therefore, is an appropriate and vital economic concern. The management task shifts from attempting to maximize wealth or incomes obtained from ecosystems to a constrained maximization based on maintaining a minimally acceptable level of ecosystem health and integrity. This changes the management focus from adapting ecosystems for human use to adapting human use for maintaining ecosystem health — which is a dramatic change in mind set.

✓ **Ecosystem management is as much management of human activities as it is management of ecological systems**. The objective is to improve human adaptability to constraints and opportunities that may be associated with preserving ecosystem health.

✓ **Narrow economic values, such as those of marketed goods and services, are only a part of ecosystems' values to human economies** (Fig. 1). Economic values of ecosystem services include direct user values, such as timber, as well as indirect values, such as flood moderation of forests and biodiversity. Many of these values are not monetarily quantifiable. Economists have developed an array of monetary valuation techniques, based on willingness to pay or to be compensated, to place monetary values on some ecosystem services, but not all services can be valued in these contexts.

✓ **The valuation of ecosystems range from aggregations of narrow individual valuations to broad social valuations reflecting moral and cultural factors**. While the former have some reliable quantifiability, the latter are unlikely to be quantified, although they have important roles to play in

Fig. 1. Some values (e.g. timber) obtained from ecosystems are readily measured and priced. The value of many others (e.g. carbon storage and flood control) are only beginning to be measured. (Photo courtesy of U.S.D.A. Forest Service.)

management decisions. While individualistic direct or indirect use valuations can be quantified in some reasonably objective basis, the broad social values emerge only through community dialogue. Ecosystem management requires the structuring of decision processes to allow the emergence of these social values. These social values are a result of collaborative "visioning" processes, which is an integral part of ecosystem management.

✓ **Private ownership and management of ecosystem resources can provide reasonable stewardship of the flow of marketable products from ecosystems.** However, this management regime does not always lead to the highest and best use of ecosystems when there are significant spillovers to the public. This is often the case when ecosystems provide a variety of public services, such as flood control, nutrient cycling, groundwater recharge, and biodiversity (Fig. 2). Unfortunately, public ownership and

Fig. 2. Wetlands, such as these salt water wetlands along the Chesapeake Bay, are sought after development sites in part because of their close proximity to scenic waterways. This private value is usually in conflict with the substantial public values provided by the wetland's ecosystem services. (Photo courtesy of U.S. Fish and Wildlife Service.)

management does not necessarily lead to the highest and best use or even better use than a private regime. Poor public management can arise from ignorance about ecosystems or social values.

✓ **Instruments of ecosystem management are broad, ranging from rigid, centralized prescriptions to flexible, decentralized economic incentives**. Their selection can significantly affect the efficiency of management. Examples of flexible instruments include negotiations between stakeholders, liability and pricing, and market emulation. Economically efficient resource management requires that price-based instruments for resource access be based on a full assessment of the costs associated with that access rather than upon revenue generation.

✓ **Ignorance about how ecosystems function and about preferences of future societies suggest strong burdens of proof are needed for substantial interventions in those ecosystems**. This is a precautionary approach to ecosystem use. Unfortunately, no well-defined indicators of health and integrity exist that are easy to apply. Most likely, health and integrity will be based on a combination of human values and "scientific" indicators.

PRACTICAL CONSIDERATIONS

Ecological economics recognizes that there are physical laws that constrain economic possibilities, particularly unconditional, unlimited economic growth. Economies are dependent on the ecosystems in which they are embedded. Ecosystem and economic management models must shift thinking toward maintaining natural capital and its flow of services to sustain even contemporaneous levels of economic activity. Full system management includes managing both ecosystems and economies. Preserving the health and integrity of ecosystems is a dominant goal emerging from this field. While ecological economics recognizes biophysical limits to human economic activity, it recognizes that adapting the human economy to ecological constraints will be shaped by human choices and preferences (Box 2).

What is critical knowledge under an ecological economic stewardship paradigm? The ecological economic stewardship framework requires scientific knowledge of both how natural ecosystems respond to economic activity, as well as how economic activity responds to ecosystem changes. An analytical construct at this boundary is a full ecological–economic, input–output matrix. In this analysis, one quantifies flows of material, energy, nutrients, etc., between the economic and ecological systems and establishes how one system affects the other. Economists have developed such a model for the economy alone, and ecologists have established energy flow models for ecosystems. However, little progress has been made in coupling these two separate models in any meaningful practical way. Researchers at the University of Maryland, for example, are currently undertaking a potentially useful coupling. These researchers have developed an ecosystem model of the Patuxent, Maryland watershed, where flows of nutrients and energy flow between spatial cells. Economic land uses are predicted, with the ecosystem configuration being an input to that prediction. Land use then feeds back to the ecosystem through runoffs based on land use. The system, which is dynamic, can be used to predict land use and ecosystem configuration. This modeling of the

Box 2
Important Questions When Using Ecological Economics

- What does society wish to become and what does it value?
- What is the requisite health of an ecosystem relative to that social objective?
- What set of human economic artifacts, structures, and processes is feasible within that requisite healthy ecosystem?
- How can society use the adaptability of human economies to assure they meet their own welfare needs as well as the needs for preservation of a healthy ecosystem?

ecologic–economic interaction is useful for foreseeing implications of management decisions, and thereby valuing different options. Decision makers and other can use these implications as information in collaborative decision settings.

Cost–benefit analysis is frequently used as a social norm for management decisions. The practical implementation of cost–benefit analysis as a management method raises justifiable concern. A practical version of cost–benefit analysis uses money as the metric for measuring costs and benefits of changes in ecosystem resource use. Researchers have developed a variety of techniques to establish this monetary metric. Although monetization of values associated with marketed products and services of ecosystems appears straightforward, many ecosystem values cannot be ascertained directly by observing market behaviors. In response, economists have attempted to take monetary valuation into non-traditional areas. Non-quantifiable values are just as important as narrow, quantifiable ones; yet, it is easy for people to disregard them in the management process.

Bottom-line net benefits or costs are not enough to make the hard management decisions. All decisions will affect differently the various stakeholder groups. These decisions will ultimately be political in nature, requiring an understanding of exactly who gains and who loses. Decision-based cost–benefit analysis must maintain an account of gains and losses by stakeholder groups. For example, a decision about a forest cut might pit recreational groups against local loggers and timber firms. These groups may differ geographically, and even temporally if recreational use is sustainable and logging is not. Ecosystem management problems often involve geographically widespread and diverse benefits, but highly localized costs. Knowing the magnitudes of the gains and losses of stakeholders can help decision makers in weighing the equity issues.

An economic analysis that explores the impacts of ecosystem use must always address the costs of such impacts. Knowing that a particular ecosystem use will result in job and income enhancements is only half the picture. The other half is how many jobs and how much income would be sacrificed as a result of that use.

When there is an objective of using the ecosystem in the "highest and best" manner, there must be some type of valuation system for making the inevitable trade-off decisions in ecosystem management. Traditional economics offers a potentially useful array of valuation methods, all directed toward determining monetary values for various uses. These valuation procedures seek measures of what individual members of society would be "willing to pay" to have more ecosystem services of a certain type, such as recreation, or what they would be "willing to accept" in compensation for denial of these services. An entire array of values, from direct use of a resource, such as timber harvest, to non-use, such as cultural values, are absolutely necessary for establishing a complete picture of ecosystem service values. These methods include both observed and hypothetical techniques. Although these methods do provide meaningful clues as to the relative values of different ecosystem uses, one must remember that monetary valuations are not the only types of values. Moreover, individualistic valuations may differ from a more socially oriented valuation. Nevertheless, these traditional economic valuation methods help paint a picture of value that is superior to simply counting trees logged or livestock grazed.

Valuation is a critical task of ecosystem management. Social welfare norms are the basis for valuations. These norms are sometimes explicit and carefully specified in such institutions as statutes and their resulting regulations. For example, banning ozone-depleting chemicals is an explicit statement of social norms. Norms become explicit through judicial decisions, whether they are interpreting intent or administering common law. However, it is often the case that social norms and values are poorly specified, or only specified on a case-by-case basis when circumstances arise where conflicting values are at stake. For example, the Endangered Species Act appeared to establish well-defined values for every species; the value of each species is potentially infinite when it is near extinction. The implications of this valuation have set in motion many attempts to change the values implied for species. When norms are poorly specified, managers face difficulties because they have no well-defined metric to measure efficiency or inefficiency in management. This often results in managers trying to understand implied values through marginal decisions, then awaiting political repercussions. Later society reformulates the implied value system after it realizes the implications of prior decisions.

There are many economic techniques for valuing ecosystem services. However, these techniques are sometimes costly and cumbersome to implement in particular situations. Short-cut techniques are available and managers may be able to transfer information from other ecosystem contexts to the one of concern to them.

Ecosystem management is a complex task. It involves not only managing complex natural capital systems, but it should also involve managing complex human economies. Certainly, decision makers must evaluate full valuation of uses, both direct and indirect, when gauging the trade-offs involved in decisions. In addition to asking about the adaptability of the ecosystem to changing uses, the ecosystem manager must ask about the adaptability of the human economies connected to and dependent on these natural systems. An implication is that social-based valuations, such as those arrived at through stakeholders meetings and other means may be superior to individualistic, narrowly economic valuations.

SELECTED READING

Bockstael, N., R. Costanza, I. Strand, W. Boynton, K. Bell, and L. Wainger. 1995. Ecological economic modeling and valuation of ecosystems. *Ecological Economics* 14(2): 143–161.

Costanza, R. (ed.). 1991. *Ecological Economics: The Science and Management of Sustainability.* Columbia University Press, New York.

Cummings, R.G., D.S. Brookshire, and W.D. Schulze. 1986. *Valuing Environmental Goods: An Assessment of the Contingent Valuation Method.* Rowman and Allanheld, Totowa, NJ.

Daily, G. 1997. *Nature's Services: Societal Dependence on Natural Ecosystems.* Island Press, Washington, DC.

Daly, H.E., and J.B. Cobb, Jr. 1989. *For the Common Good: Redirecting the Economy Toward Community, the Environment, and a Sustainable Future.* Beacon Press, Boston.

Freeman, A.M. III, 1994. *The Measurement of Environmental and Resource Values: Theory and Methods.* Resources for the Future, Washington, DC.

Goudie, A. 1994. *The Human Impact on the Natural Environment.* MIT Press, Cambridge.

Jansson, A.M., M. Hammer, C. Folke, and R. Costanza (eds.), 1994. *Investing in Natural Capital: The Ecological Economics Approach to Sustainability.* Island Press, Washington, DC.

Loomis, J.B. 1993. *Integrated Public Lands Management: Principles and Applications to National Forests, Parks, Wildlife Refuges and BLM Lands.* Columbia University Press, New York.

Mitchell, R.C., and R.T. Carlson. 1989. *Using Surveys To Value Public Goods: The Contingent Valuation Method.* Resources for the Future, Washington, DC.

Uncertainty and Risk Assessment

Why Risk Assessment Is Important To Ecological Stewardship

Wittingly or not, we accept some level of risk in everything we do. In ecosystem management, the goal is to manage a system of natural resources and ecological processes so that the overall risks from both natural disturbances and human influences are held to acceptable levels at minimum cost (Box 1). A few examples of risks that affect management include: changes in the demand or prices of timber, effects of forest policies on employment, impact of additional roads on the viability of wildlife populations, and ability of increased riparian reserves to improve spawning habitat for endangered fish. Some of these risks will be more important than others to the overall health of the ecosystem. The purpose of a risk assessment is to provide better information for making decisions about what risks to manage and how to address them (risk management).

This summary was drafted by Andrew Malk with contributions from David Cleaves and Richard Haynes. It is based on the following chapters in Ecological Stewardship: A Common Reference for Ecosystem Management, Vol. III:

Cleaves, D. and R. Haynes. "Risk Management for Ecological Stewardship."

Haynes, R. and D. Cleaves. "Uncertainty, Risk, and Ecosystem Management."

KEYWORDS: Fire suppression, prescribed fire, uncertainty, risk management, probability

Risk assessments can be used to: (1) compare risks in the same ecosystem or management option; (2) clarify otherwise fuzzy notions about types and relative severity of risks; (3) provide insight about the interactions between management options and the ecosystem; (4) focus public debate on the most significant risks; (5) point out weaknesses in the knowledge base to prioritize research and the collection of information; (6) display tradeoffs among risks and between risks and benefits and costs of management options; (7) educate the public about the inner workings of the system; (8) provide input in the design of ecological monitoring, and; (9) provide a basis for quantifying and describing the risk attitudes and behavior of ecosystem managers.

KEY FINDINGS

✓ **Uncertainty in any ecosystem process or function can pose both risks (probabilistic loss) and opportunities (probabilistic gains).** Ecosystem management processes must balance risks, opportunities, and their relative likelihoods. In doing so, managers must understand the different perceptions and values about risk held by the public.

✓ **Ecosystem risk management is a continuous process.** Risks change over time so risk management involves continuous manipulation of multiple risks to the physical, biological, and human components of an ecosystem by selecting programs and projects to hold the overall risk to acceptable levels at minimum cost. Risk management decisions involve finding ways to reduce probabilities, lower the losses, interrupt exposure pathways, and collect information to better predict events (Fig. 1).

✓ **Flexibility is a key component of risk management.** Because risk policies shape managers' attitudes toward accepting risk, managers must have the flexibility to adjust to changes in the natural, business, or political environment. The landscape perspective of ecosystem management can allow different risks to be managed at the scale most consistent with their exposure and with the span of discretion available to the decision maker.

✓ **While risk has a strong probabilistic component and follows basic laws of math, public perceptions of risk reflect factors more complex than mere probability.** Many individuals, decision makers, and organizations have difficulty thinking in probabilistic terms. They may demand scientific certainty even when it is not attainable, and discount evidence that does not agree with strongly held beliefs. People generally fear uncontrollable risks far more than controllable risks, and risks that could result in catastrophe more than those with less drastic consequences, even if the latter are more probable. The media, which are often the principal source of public information about risk, report the most vivid and catastrophic events, and focus on the most disturbing or stimulating elements for "human interest" value. The media operate under strong deadline constraints, so they rely on "data" selected for availability rather than accuracy, and may select experts to describe risks because of their willingness to talk rather than their abilities to evaluate and predict system change.

✓ **Land managers seem typically risk-averse.** They tend to manage risks by meeting the standards rather than evaluating the risks associated with tradeoffs. Top-down guidelines, regulatory standards, and Best Management Practices frequently substitute for an informed assessment of risk.

Box 1
Definitions

Risk is exposure to a chance of loss. A full characterization of a risk must include an estimate of the magnitude of possible loss, the chances, and the exposure. An example would be "a five percent chance of fusiform rust hitting a loblolly pine plantation within the next 10 years and causing a loss of 10 percent in growth and yield".

Acceptable risk is an expression of "how safe is safe enough".

Ecosystem risk assessment is a process to evaluate the likelihood that adverse biophysical or social/economic effects may occur as a result of an event or management action.

Risk management is the continuous process of allocating an organization's or society's scarce resources to reduce or maintain important risks at acceptable levels.

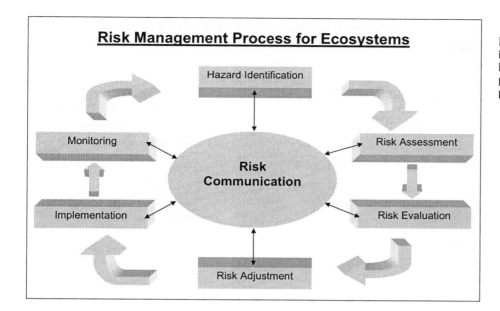

Risk Management Process for Ecosystems

Fig. 1. Risk management is an iterative cycle of evaluating information, adjusting management practices, and updating the risk profile.

PRACTICAL CONSIDERATIONS

The US Environmental Protection Agency (EPA) has developed an ecological risk assessment framework that lays out three clear steps for initiating an assessment. The first step is formulating the problem, which involves identifying the nature and array of management decisions, identifying and specifying the elements of the system, and describing the desired futures. The second step is risk analysis, which characterizes how a disturbance (natural or management-induced) interacts with the biophysical, social, and economic elements of an ecosystem, and how it causes adverse effects and under what circumstances. The third step is risk characterization, where interactions of stressors with elements of an ecosystem are interpreted in terms of the likelihood of various levels of effects.

The Forest Ecosystem Management and Assessment Team (FEMAT) assessment of management options for the Pacific Northwest's forests demonstrated that useful assessments of uncertainty depend on systematic assessment procedures and careful attention to how they will be communicated. Their experience offered the following lessons:

- Help both decision-maker and expert understand each other's perspective (Box 2);

- Train both experts and decision-makers in how to use uncertainty expressions and how to recognize high quality judgements;

- Identify vivid events, scenarios, or past decisions that could bias expert's judgements or decision-makers perceptions of the assessment results, and;

- Do not drive for consensus in group assessments. Try to develop a complete understanding of the problem though full and open discussion.

Reducing one risk can reduce or exacerbate other risks (Box 3). Reducing the risk of catastrophic fire through increased prescribed burning increases the risk of health problems and economic disruptions from prescribed burn smoke (Fig. 2). Risk reduction can also open or foreclose opportunities. The reduction of wildfire risk by aggressive fire control may allow a private owner the potential to reforest and capture investment returns. A risk standard set too tight (low)

Box 2
What Makes a High Quality Decision?

Human influences on the ecosystem result from human decision-making processes. Most decision scientists agree that a high-quality decision (1) accurately describes the problem and the criteria for solving it, (2) uses available information, (3) generates and chooses from a wide range of alternatives, (4) distinguishes facts, myths, values, and unknowns, (5) describes consequences associated with alternative problem solutions, and (6) leads to choices that are consistent with personal, organizational, stakeholder, or other important values.

Fig. 2. Not everyone perceives the
same risks and opportunities. For
example, the natural resources
manager "burner" is likely to
perceive risks and opportunities of
prescribed fire differently from
society.

PRESCRIBED BURNING:RISKS AND OPPORTUNITIES

	"BURNER"	OTHERS (Society)
Benefits and Opportunities	• Regeneration • Vegetation Control • Hazard Reduction • Animal Control • Disease Prevention	• Hazard Reduction • Avoid Larger Wildfire Emissions • Improved Wildlife Habitat • Improved Forest Health
Risks and Costs	• Contracting • Compliance • Tree Injury • Overhead Costs	• Human Health • Highway Accidents • Visibility Aesthetics • Escaped Fires

Fig. 2. Not everyone perceives the same risks and opportunities. For example, the natural resources manager "burner" is likely to perceive risks and opportunities of prescribed fire differently from society.

or too loose (high) can allow other risks to reach destructive levels or create nonlinear cumulative impacts. For example, much of the thinking about biodiversity protection pushes for preservation of large systems of old-growth forest reserves, with a minimum percentage of area in connected, old growth condition. However, following this policy can create risks of wildfire conflagration in continuous stands of accumulated fuels, and forest health problems in overmature, overstressed trees.

The degree of risk acceptable to a diverse public is difficult to estimate. Different parts of society may view stress, events, and management actions at different spatial

---

Box 3
Evaluating Risk Tradeoffs

Reintroduction of fire is a well-established goal in the minds of ecosystem managers, but the level of activity is still less than half that desired. Barriers to increased burning include funding, public opinion in urban interface areas, liability risk, conflicting agency risk-taking policies, and the complexities of dealing with often conflicting air quality and environmental laws. The underlying cause is that burning offers a complex array of tradeoffs among benefits and costs, opportunities and risks. Efforts to realize one or more of these objectives often involve trading off accomplishments of others.

Standards for species protection represent one such tradeoff. Burns with unacceptably high risks of escaping and damaging species or habitats may be postponed or canceled. It is unclear how these risks are accounted for in burn planning and decision-making. Snag retention standards have become issues not only because snags must be afforded protection from fire damage, but also because they pose a hazard to fire crew safety and can serve as an ignition source.

To assist in making tradeoffs such as these, federal agencies have developed several initiatives to improve the capabilities of the organization in risk-based planning for prescribed burning. A key ingredient is the collaborative process to ascertain the relative values and create innovative solutions to localized situations. Within the Forest Service, fire managers are approaching the fuels management question with a systematic process called Fire Loss Management that requires teams of fire management, other professionals, and the public to rate sections of forests on probability of ignition, level of hazard, and value-at-risk. Attention is focused on one element at a time so different perceptions of values can be fully explored. Aggregating these assessments into a comparative risk assessment allows better allocations of effort across prevention, detection, loss reduction, initial attack, and other elements of the strategy.

scales. For example, the risks of human community dysfunction due to old growth reserves in western Oregon are viewed with a much different perspective by residents of those communities than by urban citizens on the East Coast. However, methods exist to help disclose risk attitudes of individuals or small groups of decision-makers.

Risk policy is composed of rules, standards, and other instruments that signal what risks are most important and what levels of those risks are acceptable. Protection standards are written to assure particular behaviors of managers toward ecosystem components. An enduring standard in wildland fire fighting has been the "10 a.m." fire suppression standard, mandating full control on every fire during the first day, or by the beginning of the next burning period. Although this "standard" has driven the thinking and planning of fire-fighting agencies since the 1930s, this "fire exclusion" policy is now being revised in light of growing recognition of fire's ecological benefits and the long-term accumulation of hazardous fuels. The agency's risk policy has been modified to allow more discretion on the part of the fire manager, but the suppression standard is deeply ingrained in the expectations of policy makers and the reward systems of managers.

SELECTED READING

Bishop, R.C. 1979. Endangered species and the safe minimum standard. *American Journal of Agricultural Economics* 61: 376–79.

Cleaves, D.A. 1994. Assessing Uncertainty in Expert Judgments about Natural Resources. USDA Forest Service General Technical Report SO-110. Southern Forest Experiment Station, New Orleans, LA.

Fischhoff, B. 1985. Managing risk perceptions. *Issues in Science and Technology* 2(1): 84–86.

Kleindorfer, P.R., H.C. Kunreuther, and P.J.H. Schoemaker. 1993. *Decision Sciences: An Integrative Perspective.* Cambridge University Press, New York.

MacCrimmon, K.R., and D.A. Wehrung. 1986. *Taking Risks: The Management of Uncertainty.* The Free Press, New York.

Marcot, B.G., and H. Salwasser. 1991. Views on Risk Analysis for Wildlife Planning and Management in the USDA Forest Service. Presented at Society for Risk Analysis Annual Meeting, 8–11 Dec. 1991, Baltimore, MD.

Slovic, P. 1987. Perception of risk. *Science* 236: 280–285.

Suter, G.W. (ed.). 1993. *Ecological Risk Assessment.* Lewis Publishers, Chelsea, MI.

United States Environmental Protection Agency. 1992. Framework for Ecological Risk Assessment. *Risk Assessment Forum.* EPA, Washington, DC.

Wack, P. 1985. Scenarios: uncharted waters ahead. *Harvard Business Review* 85(5): 73–89.

Economic Tools For Ecological Stewardship

Why Economic Tools Are Important To Ecological Stewardship

Ecosystems are the source of diverse products and services that sustain and enhance our life. When European colonists first arrived, these products and services, such as clean water for drinking, and fisheries and forests for timber and wildlife habitat, were plentiful. As colonists settled the United States, they converted forests to homes, tools, and heat, wetlands and prairies to farms, and wildlife to food and clothing. As a result, some ecosystems are now close to disappearing altogether and others may already be gone. As they go, we in society lose the cultural and natural heritage of that region. We also lose the opportunity to benefit from the knowledge of the intricate workings and interrelationships that exist in a particular ecosystem type. With each ecosystem extinction, society eradicates both the history of the place and its future.

This summary was drafted by Robert Szaro with contributions from Paige Brown. It is based on the following chapter in Ecological Stewardship: A Common Reference for Ecosystem Management, Vol. III:

Brown, P. "Tools for Ecological Stewardship."

KEYWORDS: Easements, economic incentives, property rights, taxes, policy

---

Box 1
Property Rights Tools

- Conservation Easements
- Covenants
- Deed Restrictions
- Conservation Agreements
- Land Donations
- Land Exchanges
- Land Trusts

As resources continue to disappear and become degraded, researchers, policy makers, and others developed means to halt these losses. The tools, goals, and approaches to halting destruction of natural resources have changed as knowledge has increased and threats to natural resources have changed. Some of the first laws regulating natural resource use attempted to deal with the devastating effects of hunting. In 1973, the Congress enacted the Endangered Species Act, which prohibits the "taking" of plants and animals threatened with extinction. What "taking" means and how it relates to private property is still being debated, as is evidenced by the "Sweet Home" lawsuit, the Spotted Owl versus logging debate in the Pacific Northwest, and the gnatcatcher versus housing developments in Southern California.

These conflicts illustrate that relying solely on regulatory methods misses key conservation opportunities, and may engender unnecessary opposition to habitat protection. However, each tool must operate within a regulatory framework to ensure that minimum habitat is maintained, contracts are enforced, and promises are kept. It is critical to seek creative and fair methods to ensure that we all help share the burden for supporting habitat protection, as we all take part in some of the benefits.

KEY FINDINGS

✓ **Property rights tools** are actions that alter the definitions of property use or ownership (Box 1). The tools are voluntary, not regulatory, and do not necessarily involve financial transactions. These tools are based on the U.S. legal concept of property rights, which is that property rights are essentially a bundle of rights that are divisible from each other. At one end of the spectrum lies property donation, which amounts to donating the entire "bundle," i.e., donating the land itself. The other property rights tools — covenants, deed restrictions, and conservation easements — involve separating out some of the property rights, such as the right to certain uses. Land Trusts use these and other tools to protect the rural or environmental character of land. These tools may prohibit development, or may require certain actions, such as development of a management plan (Kusler and Opheim 1996).

✓ **Taxes** are one of the most powerful market-based policy tools for providing incentives or disincentives for encouraging improved ecological stewardship (Box 2). Income tax and property tax deductions are two means for providing incentives for conservation activities such as instituting conservation easements, donating land, habitat enhancement, and restoration activities. Taxpayers can reduce estate or inheritance taxes by putting land under a conservation easement. Such taxes still may encourage development because the heirs may be land-rich, but cash-poor, and unable to pay the taxes without selling or developing the land.

✓ **Incentive-based policy tools**, according to some economists, are an efficient way to incorporate environmental concerns into the market. First, an environmental goal is defined, such as reducing a pollutant or improving forest management, and then, a system of monitoring or trading is designed so that those affected parties choose the manner in which they reach the goal. This system provides flexibility in how to reach the desired goal. Incentive-based policy tools use information to create incentives to change producer and consumer behavior. Moreover, because the programs, such as

---

Box 2
Tax Policies

- Income Tax Deductions
- Property Tax Reductions
- Property Tax Incentives
- Estate Taxes

Fig. 1. An example of an incentive tool is a fund established by Defenders of Wildlife to compensate ranchers for livestock losses caused by wolves in the Northern Rockies. (Photo courtesy of U.S. Fish and Wildlife Service.)

wolf reintroduction or timber certification, are voluntary, participants are generally less resistant (Fig. 1). These programs give value to important non-market benefits such as biodiversity, clean air, and clean water that landowners or managers can potentially capture (Box 3).

✓ **Private–public partnerships**: These mechanisms generally require involving both the private and public sector, often some form of governmental oversight, and a regulatory framework within which to operate. Some tools may be nested in others; for example, Safe Harbor Agreements may be part of a habitat conservation plan. Many of these tools are more complicated than other tools because they often involve extensive conversations among concerned parties and require a great deal of information to determine

---

Box 3
Examples of Incentive-based Tools

- User Fees
- Ecolabeling: Forest, Marine, and Agricultural Products Certification
- Forest Stewardship Council
- Green Credit Cards
- Green Investments
- Partnerships for Sustainable Development in the Pacific Northwest
- Biodiversity Trust Funds
- Contracts for Carbon Sequestration
- Carbon Sequestration Projects
- Benefit Sharing for Genetic Resources

Box 4
Examples of Private–Public Partnerships

- Habitat Conservation Plans
- Habitat Conservation Plan for the Red-Cockaded Woodpecker
- Assurances or "No Surprises"
- Tradable Development Permits
- Habitat Transaction Methods
- Mitigation Banking
- Conservation Banking
- Adopt-An-Animal/Habitat Programs
- Community-Based Partnerships
- Conservation Partnerships
- Community Re-investment Act
- Corporate Conservation Services Incentive Programs

---

which lands should be designated as development or conservation zones. These partnership efforts are generally more complex than those that allow a single private landowner to make a decision affecting only his or her land, such as a conservation easement (Box 4).

✓ **The Government**, at the federal, state, and local levels, has a critical, and occasionally contradictory, effect on habitat through enforcement of regulations that prevent actions that harm species or neighboring property, incentive and disincentive programs, infrastructure projects, tax and land use policies, and water management and ownership. Eliminating perverse subsidies that encourage environmental degradation significantly affects the environment and is fiscally prudent. Examples include flood insurance in flood-prone, environmentally sensitive areas, unwarranted infrastructure projects whose benefits do not exceed their costs, such as some dams and highways, and agricultural subsidies that encourage crops in inappropriate areas where they may require excess chemical applications or cause erosion (Box 5). Conversely, the government is a key player in virtually all types of tools that improve ecological stewardship. Government programs may offer direct payments through programs such as the Wetlands Reserve Program, which pays for the protection and restoration of wetlands (Fig. 2). Many programs also offer technical assistance; for example, the Forest Stewardship Program helps landowners manage for multiple benefits.

Box 5
Government Programs and Incentives

- Bonds
- Key Federal Programs
- Key State Programs
- Dedicated Trust Funds
- Earmarked Tax Revenues
- Earmarked Fees
- Special Assessments
- Exactions or Proffers
- EPA Grants
- Add Requirements to Existing Programs
- Affinity Merchandise

Fig. 2. Since 1935, Federal Duck Stamps have generated $501 million that has been used to preserve 4.4 million acres of waterfowl habitat in the United States.

First "Duck Stamp" 1934–1935. Migratory Bird Hunting Stamp. Design by Jay Norwood "Ding" Darling.

✓ **Voluntary initiatives**: "Conservation begins with an individual's decision" (USDA 1996). Property rights tools, tax policies, partnerships, and governmental programs are all critical tools to encourage ecological stewardship. In large part, however, the ultimate success of most tools depends on the desire of those involved to have a positive effect. There are many instances where groups and individuals from the private and public sector have initiated improved management, preservation, restoration, and conservation activities simply because they value our natural heritage

PRACTICAL CONSIDERATIONS

Clearly, the United States is using a broader range of tools for encouraging ecological stewardship than early hunting restrictions. Americans are finding new ways of using existing property rights tools, government programs, and tax policies to achieve conservation ends. Also, Americans are trying new, more flexible and comprehensive approaches within existing regulations, such as the Endangered Species Act. Simultaneously, people are recognizing such ecosystem values as water purification, carbon sequestration, and the potential for new medicines. Consumers will be better able to express their preferences for products from well-managed lands once ecolabeling schemes become more widely recognized and implemented. By the same token, ranchers, farmers, and forest managers, whose products are based on natural resources, will be better able to shift to sustainable management practices if they are able to realize the benefits.

Many of the newer tools and examples stress partnerships (Box 4). These partnerships offer the opportunity to create an integrated stewardship plan across a wider area. By working on a broader landscape, concerned parties can share the burden of ecosystem planning and make decisions in a more strategic fashion. Additionally, if the planning process includes the full range of stakeholders, there may be less resistance to including ecosystem concerns into land-use planning, particularly if additional regulation becomes un-necessary. Although such partnerships are not always easy or successful, nor do they always reduce the need for regulations, they do offer opportunities for comprehensive planning and dialogue between groups with different expertise. Both of these qualities potentially enhance environmental benefits.

There is an increasing trend towards being proactive, rather than reactive. Some efforts use incentives to improve stewardship, such as recent forest certification efforts. Certification can provide a means for forest managers to maintain the land as standing forest and practice good stewardship. Similarly, environmentalists and others have criticized and challenged grazing on public lands. Rancher Jim Winder's response was to investigate whether he could alter his management practices such that his livestock could coexist with native range species. Can ranchers and environmentalists really work together for a common solution, rather than being at odds? Both certification and improved management could help circumvent the need for governmental regulations. Also, when voluntary efforts go beyond new laws, as was the case with Seven Islands, a timber company in Maine, the effect of government regulations is lessened. ShoreBank Pacific offers yet another example of proactive efforts. ShoreBank Pacific in southwest Washington state not only screens out investments with negative environmental impacts, but nurtures positive investment opportunities by partnering with local groups.

There is an overall trend towards experimentation. Experimentation is essential if society is to solve the problem of habitat loss and degradation without causing undue economic burden. The overarching lesson is that virtually everyone can contribute to improving ecological stewardship, provided they understand the array of tools available to them and realize what will be lost if they do not use them.

SELECTED READING

Bean, M.J., and D.S. Wilcove. 1997. The private-land problem. *Conservation Biology* 11(1): 1–2.
Bowles, I.A, D. Clark, D. Downes, and M. Guerin-McManus. 1996. *Encouraging Private Sector Support for Biodiversity Conservation.* Conservation International Policy Paper. Vol. 1.
Clark, D., and D. Downes. 1995. *What Price Biodiversity?* Center for International Environmental Law, Washington, DC.
Environmental Protection Agency. 1997. *Environment Finance Program: A Guidebook of Financial Tools.* EPA.
Hudson, W.E. (ed.). 1993. *Building Economic Incentives into the Endangered Species Act.* Defenders of Wildlife, Washington, DC.
Keystone Center. 1996. *The Keystone National Policy Dialogue on Ecosystem Management.* Keystone Center, CO.

Kusler, J. and T. Opheim. 1996. *Our National Wetland Heritage: A Protection Guide.* Environmental Law Institute, Washington, DC.

Land Trust Alliance. 1997. *Partnership Notebook: a Land Trust Guide to Federal Agriculture Programs.* United States Department of Agriculture, Washington, DC.

Minette, M. and T. Cullinan. 1997. *A Citizen's Guide to Habitat Conservation Plans.* National Audubon Society, Washington, DC.

Noss, R.F., E. T. LaRoe III, and J. M. Scott. 1995. Endangered Ecosystems of the United States: A Preliminary Assessment of Loss and Degradation. USDI Biology Report. No. 28.

Information and Data Management

◆ *Adaptive Management*

◆ *Assessment Methods*

◆ *Monitoring and Evaluation*

◆ *Data Collection, Management and Inventory*

◆ *Decision Support Systems*

Adaptive Management

Why More Effective Learning And Adapting Are Important To Ecological Stewardship

Daniel J. Boorstin, the noted historian and former Librarian of Congress, opens his account of European explorers by observing: "The great obstacle to discovering the shape of the earth, the continents, and the ocean was not ignorance but the illusion of knowledge." The illusion of knowledge has also plagued natural resources management. Even as flaws become apparent, we cling to paradigms that have outlived their usefulness. Problems are dismissed as minor, or unusual, or isolated and we go back to relying on the research, concepts, and management conventions we have gotten used to. For example, long after Aldo Leopold and other researchers knew better, predators were still being treated as destructive forces to be eliminated from ecosystems. When a single approach to forestry was almost universally applied — as the staggered-setting clearcut and slash-burn approach was in the Pacific Northwest over the last 40 years — little biophysical, economic, or social knowledge was being sought about alternative approaches to management. Only after decades of mounting problems do we begin to distrust what we think we know.

This summary was drafted by Andrew Malk with contributions from Bernard Bormann and Jerry Magee. It is based on the following chapter in Ecological Stewardship: A Common Reference for Ecosystem Management, Vol. III:

Bormann, B.T., J.R. Martin, F.H. Wagner, G. Wood, J. Alegria, P.G. Cunningham, M.H. Brooks, P. Friesema, J. Berg, and J. Henshaw. "Adaptive Management."

KEYWORDS: Experimentation, monitoring, learning, adaptation, knowledge, citizens

Fig. 1. Do we have a fear of learning? (CALVIN AND HOBBES © 1992 Watterson. Reprinted with permission of UNIVERSAL PRESS SYNDICATE. All rights reserved.)

The danger of untested knowledge is no less a problem for ecosystem approaches to natural resources management. To increase the chances that ecosystems can be sustained, society needs to understand human and ecological processes, particularly agents of social and ecological change and their interactions. This understanding requires knowing more about societal values, ecological capacity, and their interactions than researchers know today (Fig. 1).

Researchers need to accelerate their knowledge and understanding of these processes because environmental and social changes are occurring rapidly as growing populations and escalating standards of living conspire to drive ever higher demands for raw materials. Society may not have much more time to learn the limits of the world ecosystem and how to live at a sustainable pace within those limits). Because of these forces, natural resource management today is much more complex than believed only a few decades ago. Today, resource managers are responsible for a mix of human-influenced "natural" ecosystems and manipulated systems in a world concerned about global climate change, biodiversity conservation, maintenance of site productivity. Their management efforts are under the scrutiny of an increasingly critical citizenry.

KEY FINDINGS

Adaptive management, itself, is an evolving concept with multiple sources of origin. Nevertheless, research and experience have shown that several general concepts are important to maximize the benefits of an adaptive approach to management. The concepts include the following:

✓ **Adaptive management is an approach to managing complex natural systems that builds on learning**. Adaptation is a positive response to change. There are a variety of ways in which the term "adaptive management" is defined and used, but learning should be at the center of the concept. Learning can be based on common sense, experience, experimenting, and monitoring. Learning, however, has to be used to adjust practices if it is to be an effective tool in natural resources management. Adaptive management should focus on accelerating learning and adapting through partnerships based on finding common ground where managers, scientists, and citizens can try to learn together to create and maintain sustainable ecosystems that can support human needs indefinitely.

✓ **Don't wait for a cookbook for adaptive management.** People learn and adapt in many ways, and the process of learning and adapting must also evolve over time. A range of learning and adapting strategies can be combined into effective solutions for the great diversity of ecosystems and possible manager-scientist-citizen partnerships. Adaptive management is a participatory approach to research and natural resources management, not the sole province of resource management professionals and highly trained experts (Box 1).

Box 1
Basic Assumptions for Adaptive Management

• Managers, scientists, and citizens need faster and more effective approaches to learning that can easily be incorporated into their everyday lives.

• Citizens and managers have as much or more at stake in learning as scientists do, and scientists and citizens can help managers adapt to changing values and information.

• Knowledge, the product of learning, must be considered as a resource of equal or greater value than the physical resources that traditionally have been the focus of management.

• The managed landscape itself contains important information — including opportunities for retrospective studies of past management and natural events — that can be given value and managed to produce knowledge for future decisions.

• A shift is needed away from the unidirectional transfer of technology from scientists to society towards better communication of new information, knowledge, and technology between citizens, scientists, and managers.

✓ **Society can no longer afford reactive approaches to learning.** Effective learning and adapting must be central to the mission of management, research, regulatory agencies, and society as a whole. Agencies must fully institutionalize the concepts of adaptive management. For example, the U.S. Forest Service mission should be defined as: Learning how to care for the land and serve the people. By designing management projects to produce knowledge along with other resource objectives, society can accelerate the learning process (Box 2). Because parallel approaches can compare different policies simultaneously, learning is more rapid than with a less structured approach that compares different policies sequentially.

> **Box 2**
> **Learning to Manage: Managing to Learn**
>
> The Ouachita National Forest is in Arkansas an example where learning to manage and managing to learn has become a priority for a national forest. Managers of the forest have demonstrated their commitment to learning and adapting in numerous ways, including openly acknowledging their lack of knowledge for managing whole systems sustainably, emphasizing a collaborative approach to management, establishing and demonstrating alternative approaches to vegetation management, and building strong partnerships with the science community.
>
> In 1990, the managers of the Ouachita ceased prescribing clearcutting as the standard harvest method. This policy shift sparked changes in the Ouachita's relations with the public and the scientific community. These improved relations focused on learning how to restore old growth, recover red-cockaded woodpecker populations, and renew shortleaf pine-bluestem grass ecosystems. Since 1990, more than 40 publications have illustrated the efforts of the Ouachita managers and partners toward improving knowledge, interactions, and decision making.

✓ **Planners must assume that a variety of pathways can meet a given objective**. People must design and test a range of pathways to achieve the objectives set by the current generation and provide future generations with more choices. For example, establishing side-by-side prescriptions is especially valuable.

✓ **New partnerships among citizens, managers, and scientists, which are focused on learning, are essential to achieving sustainable ecosystems** (Fig. 2) Society no longer accepts expert-based learning and decision

Fig. 2. In applying an adaptive management approach, managers need to be directly involved in monitoring , just as scientists must address research priorities identified by managers. (Photo courtesy of USDA Forest Service.)

Table 1. Historical roles of research and management and possible common ground between them; the common ground defines shared roles that promote learning and adaptation to build overlap between society's needs and wants and ecological capacity.

RESEARCH		MANAGEMENT	
Traditional roles	Common ground: shared roles to promote learning and adapting	Common ground: shared roles to promote learning and adapting	Traditional roles
Listen to managers	Directly answer their questions or ignore; avoid public input	Seek to understand and answer citizen and science questions simultaneously	Apply a full partnership model with citizens and scientists, where allowed by existing law
Focus on finding general understanding that applies to wide areas over a long time	Test general understanding at a specific time and place	Invest more in broader landscapes and over long time frames	Focus in the short-term on the land-base being managed to the exclusion of adjacent lands
Show historical bias in ecology toward "natural" ecosystems	Recognize that management is a dominant ecosystem process	Increase value of other resources, including information	Show a historical bias toward managing standards of trees, not whole forests
Study factors in isolation of other factors, usually at small scales and with extreme treatments	Integrate across disciplines, accept more complexity, and study whole ecosystem responses	Compare multiple approaches to achieving the same objective to increase learning	Track effects of a complex of practices across the managed landbase
Model processes and species independently	Focus on modeling ecosystem responses	Recognize model outputs as hypotheses not certain outcomes	Rely on economic resource, and allocation models
Maximize independence and credibility	Help design and interpret management experiments and monitoring for decision makers	Support independent monitoring and analyses of management experiments	Work as a coordinated management team

making, which segregate scientific learning from manager's activities. Creative solutions exist and they arise from interactions of diverse groups whose individual roles will have to change. New roles for citizens are needed to relate societal values to management, bring in fresh ideas, and challenge existing institutions. Without cross-translation of new information and knowledge among citizens, scientists, and managers — and without a management system than can readily integrate change — adaptation is unlikely (Table 1).

✓ **Making management more experimental is not an attempt to convert managers into researchers; rather, it seeks to use scientific learning tools to address managers' critical questions.** A manager using an experimental approach compares alternative strategies, each of which can be expected to achieve the same objective for the area, permits a complex of management practices, and uses statistical tools where possible (replication, random allocation of treatments, long-term monitoring).

✓ **Many questions are being answered in small-scale management experiments, but many important questions can only be addressed at large scales.** Because many environmental, social, and organizational dynamics cannot be measured at small scales, these dynamics can overwhelm small-scale studies. And because few examples of management experiments exist at large scales, developing them is challenging. Assessments and forest plan revisions and amendments should include learning objectives and approaches to begin effective learning at this large scale.

✓ **Decision space will expand over time**. Successful adaptation and rapid learning among citizens, managers, and scientists, are perhaps the only methods available to expand the range of alternatives for managers and society to increase the condition where societal values and ecological capacity are simultaneously met. Society expands the decision space by finding new and creative solutions, such as increasing compatibility among resource uses and developing management actions to mimic natural disturbances.

PRACTICAL CONSIDERATIONS

Many managers respond to descriptions of adaptive management concepts with the question, "How will this be different from current and past practice?" Many research scientists also see little need to change their priorities and approaches. The differences are large and numerous for both managers and scientists. Principal differences include the following:

Anticipate: don't wait for surprises. Once society accepts that future conditions cannot be predicted and controlled with great certainty, and that a single best practice cannot be determined in advance, then society can accept that it needs to learn as part of its management, especially with managers, scientists, and citizens learning together. Research should address durable problems to provide knowledge that supports future decisions.

Learn and adapt actively. Simple first steps to accelerate learning and adapting include better documentation of activities, stating anticipated outcomes in advance, comparing two or more management approaches simultaneously, and identifying trigger points to facilitate adaptation.

Include learning objectives in planning, analysis, and decision documents. Learning becomes instantly institutionalized if learning objectives, clients, and methods are identified in land management planning and project proposals as well as associated National Environmental Protection Act and decision documents. Learning at multiple geographic scales is facilitated when managers and others coordinate tiered planning and decision documents.

Manage for ecosystems. Many of the most important policy questions are more relevant as scale increases, which fundamentally changes information requirements for management and approaches to research. Managers need to recognize the importance of information coordination and accessibility, sampling to represent broad areas, and large-scale management experiments.. Scientists should study the influences of broad policies and link study of natural processes and management practices across multiple spatial and temporal scales.

Synthesize and integrate. Basic and applied sciences associated with individual resources do not provide sufficient information to achieve ecosystem sustainability. Society needs science that underpins predictions of ecosystem responses and helps managers achieve compatible production of diverse ecosystems goods and services (joint-production) more than it needs science that supports a land allocation-based approach to land management.

Managers may think implementing adaptive management means starting anew, that nothing currently being done is valid. On the contrary, managers can take may initial steps to maximize learning from their current projects, including the following:

Look for learning opportunities in current projects to demonstrate concepts. Learning is possible in every project, and modifying projects already underway is a good way to start. For example, if one has developed a single prescription for a project, he or she can find a comparable area to use as a no-management comparison and extend baseline and post-implementation monitoring to that area.

Share the concepts of adaptive management with possible partners. Look for enthusiasm and creativity. Managers, scientists, and citizens show considerable interest in the topic: what and how to learn in the normal course of managing. Managers should look for leadership from local groups, or, if unavailable, initiate these discussions themselves. They should try to take advantage of enthusiasm and creativity wherever they are lucky enough to find them.

Build a partnership for learning by creating an inclusive, safe, learning environment. Use the power of the collective knowledge and idea base. Promote courteous disagreement, it is often the source of alternative pathways to achieving the same goal. By not insisting on a single path, one can help everyone see the advantages and disadvantages of different approaches.

Explore failures and successes to learn from them. If one remembers a project that negatively affected a resource, he or she should explore what went wrong to avoid repeating the mistake.

SELECTED READING

Bella, D.A. 1987. Organizations and systemic distortions of information. *Journal of Professional Issues in Engineering* 113: 360–370.

Bormann, B.T., M.H. Brookes, E.C. Ford, A.R. Kiester, C.D. Oliver, and J.F. Weigand. 1994. *A Framework for Sustainable-Ecosystem Management.* U.S. Forest Service Pacific Northwest Research Station, Portland, OR.

Boorstin, D.J. 1983. *The Discoverers.* Random House, New York, NY.

Byerly, R., Jr., and R.A. Pielke, Jr. 1995. The changing ecology of science. *Science* 769: 1531–1532

Gunderson, L.H., C.S. Holling, and S.S. Light. 1995. *Barriers and Bridges to the Renewal of Ecosystems and Institutions.* Columbia University Press, New York, NY.

Haber, S. 1964 *Efficiency and Uplift: Scientific Management in the Progressive Era, 1980–1920.* University of Chicago Press, Chicago, IL.

Hilborn, R. 1992. Can fisheries agencies learn from experience? *Fisheries* 17(4): 6–14.

Holling, C.S. 1978 *Adaptive Environmental Assessment and Management.* John Wiley & Sons, New York, NY.

Lee, K.N. 1993. *Compass and Gyroscope: Integrating Science and Politics for the Environment.* Island Press, Washington, DC.

National Research Council. 1990. *Forestry Research. A Mandate for Change.* National Research Council, National Academy Press, Washington, DC.

Senge, P.M. 1990. *The Fifth Discipline: the Art & Practice of the Learning Organization.* Currency Doubleday, New York, NY.

Assessment Methods

Why Assessments Are Important To Ecological Stewardship

In recent years, natural resource management agencies have spent millions of dollars on natural resource assessments. Why have they made such large investments in these planning documents? The answer is at least partly because decisions about the use and regulation of natural resources have always depended on information and data. Assessments have become a primary source of data and information on resource conditions. In the past, management focused on single resources, independently assessing such resources as timber supply or endangered species habitat. Today, managers are increasingly using an ecosystem approach to manage natural resources, thus greatly expanding the scale of ecological assessments. Assessments have expanded spatially to cover large geographic scales, such as ecoregions. They provide a more comprehensive characterization of resource conditions by going beyond gathering data on amount and distribution of resources to collecting data on ecosystem processes, structure, and function. Assessments cover a range of themes, providing social and economic data together with the ecological information that is essential to an ecosystem approach to management.

This summary was drafted by Andrew Malk with contributions from Russell Graham, Gene Lessard, and Don Schwandt. It is based on the following chapters in Ecological Stewardship: A Common Reference for Ecosystem Management, Vol. III:

Graham, Russell, Theresa Jain, Richard Haynes, Jim Sanders, and David Cleaves. "Assessments for Ecological Stewardship."

Lessard, Gene, Scott Archer, John Probst, and Sandra Clark. "Understanding and Managing the Assessment Process."

KEYWORDS: regional ecosystem assessments, ecoregions, resource planning, monitoring

In addition to wider geographic scales, assessments also cover expanded temporal scales. While assessments have traditionally provided a snapshot view of the state of the resources, assessments are increasingly being conducted in the context of an adaptive management process. In the case of the recently completed Sierra Nevada Ecosystem Project or the Interior Columbia River Basin Ecosystem Assessment, dramatic changes in land use or ecological conditions indicated the need to initiate an adaptive management process by conducting an ecoregional assessment. As the first step in the adaptive management process, assessment findings should drive decisions to change management practices. It is essential to then monitor how these management changes affected the original status of the resource.

As the initiating component in adaptive management models, assessments provide the social, economic, and ecological context for decision makers and stakeholders to better evaluate the consequences of their actions on natural resources. Assessments completed at the appropriate spatial and temporal scales provide for more informed, more cost effective, and longer lasting decisions than information gathered using more limited perspectives. Furthermore, a quality assessment can provide a scientific synthesis of the social and ecological interactions in an ecosystem. This comprehensive view of resources optimizes the allocation of limited resources for diverse public demands and values. It can establish the bounds within which desired future condition can be considered.

KEY FINDINGS

In 1992, natural resource management agencies began preparing inter-disciplinary, large scale assessments in several ecoregions across the country. These have included the Pacific Northwest, Sierra Nevada, Interior Columbia River Basin, and Southern Appalachians. Interagency teams have worked to draw out the lessons learned from this first generation of assessments (Fig. 1). Some of the more important lessons include:

✓ **Assessments are not decision making documents**. While assessments provide a synthesis of information in support of resource planning and decision making, they do not identify the best or the only course of recommended action.

✓ **Assessments should precede decision making activities**. Assessments provide current data on select issues critical to decisions. Decisions made in advance of an assessment obviously will be less informed than those made with the updated and more complete information that exists after the assessment has been completed.

✓ **Assessments should be issue driven as defined by the stakeholders**. Framing the issues addressed by an assessment is the first place to involve the public (see collaboration below). The old paradigm of agencies formulating issues, assuming these are representative for the public, and developing a program to address these issues no longer works. The frequency of appeals and litigation indicates the problems of the former approach. Once the stakeholders have framed the issues, the issues should be used to define the ecosystem boundaries, temporal and spatial scales, analysis techniques, and data needs of the assessment.

Sierra Nevada Ecosystem Project
FINAL REPORT TO CONGRESS

Status of the Sierra Nevada

VOLUME II
*Assessments and Scientific Basis
for Management Options*

Wildland Resources Center Report No. 37

CENTERS FOR WATER AND WILDLAND RESOURCES
UNIVERSITY OF CALIFORNIA, DAVIS
July 1996

Fig. 1. The Sierra Nevada Ecosystem Project Assessment is one of several recent ecoregional assessments by state and federal agencies.

✓ **There may be a greater need for partnerships in assessments than other planning steps.** This depends of course on the geographic extent of assessments and the complexity of ownership patterns, but assessments usually require an array of information and skills that are found in a variety of places. Interagency partnerships are needed to ensure that assessments are more comprehensive and to avoid duplication. By providing incentives and rewards, agencies can promote research and management staff participation in assessments.

✓ **Pre-assessment planning and strong assessment management is critical.** Preliminary planning and strong management oversight are essential to keeping the assessment on time, within budget, and focused on the issues. In addition, an information management infrastructure is needed before the assessment begins to ensure data, maps, and other valuable information are maintained in useful form for future assessments and monitoring efforts.

✓ **Assessments need to describe ecosystem processes, structures, and functions at multiple spacial and temporal scales.** These scaled relationships provide context for various scales of planning and associated policy relevant decisions. Moreover, assessments should be based on ecological principles that include humans as part of the ecosystem.

✓ **To keep assessments current, information from ongoing monitoring systems should be incorporated.** This should include not only relevant resource monitoring networks, but monitoring information on the implementation, effectiveness, and validation of relevant programs and projects (Fig. 2).

In addition to all the important information that assessments produce, the assessment process itself benefits natural resource management in two important ways.

✓ **Assessments can be a powerful vehicle for promoting public involvement in resource planning.** Collaboratively conducted ecoregional assessments facilitate dialogues among managers, scientists, citizens, and diverse public and private stakeholders. Assessments provide a forum for collaboration among an array of stakeholders. If diverse stakeholders are included in the early stages of an assessment, their interests are brought closer to consensus as the assessment progresses to a decision making process. Early public involvement creates an environment for understanding, acceptance, and support for the assessment. This environment facilitates mutual learning, fosters cooperation and trust, and establishes clear expectations from both the public and the agencies as to the conduct and use of ecosystem assessments. This situation is crucial to securing greater acceptance of the assessment findings and will facilitate later decision making forums.

Fig. 2. Results from monitoring stations can build upon baseline information gathered from assessments to document ecological trends. (Photo courtesy of USDA Forest Service.)

✓ **Assessments facilitate interagency collaboration**. The tremendous research and information demands of ecoregional assessments have required interaction between scientists and managers, both within and among government agencies and universities. Many current ecoregion assessments are under the co-leadership of both scientists and managers. Pooling talents and resources is becoming ever more critical as public and private institutions have fewer resources and specialists to conduct their own narrow assessments. By bringing scientists, managers, and citizens together, assessments can better (1) address the full range of issues and consequences at hand, (2) plan at appropriate scales, and (3) avoid costly and often confusing duplication.

PRACTICAL CONSIDERATIONS

Designing the assessment prior to collecting data is the most efficient and logical way to begin. Assessment design should include the following components (Fig. 3):

Defining Problem — The first and most important step in the assessment process is to identify the issues the assessment needs to address. These issues should be well defined and stated as testable hypotheses. Well-defined questions can focus the researchers, limiting the ecosystem components and interactions evaluated to those that address the identified issues.

Delineating Assessment Boundaries — Ecoregional boundaries are usually not drawn with total certainty and vary greatly depending on the needs of the assessment. Deciding on assessment and ecoregion boundaries should be based on the assessment issues, agency needs, and the biophysical, economic, and social attributes of the region.

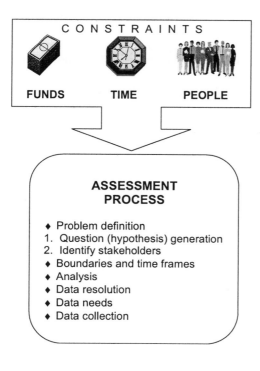

Fig. 3. Given constraints of funding, time, and expertise, it is important to define these steps early in the assessment process.

Choosing Spatial Scale — Choosing geographic scale is also based on the assessment issues and relevant ecosystem components. It is desirable for assessments to include multiple spatial scales, because it is often difficult to address adequately regional ecosystem patterns and processes that are apparent only at smaller scales. By applying various scales, assessments of large geographic areas provide context for ecosystem processes, structures, and functions that can only be characterized using smaller areas. In the end, the choice of scales will be a compromise among the people and disciplines involved in the assessment.

Selecting Indicator Variables — Indicator variables can detect ecosystem changes and should be chosen to adequately describe the ecosystem condition. If the indicator variable is to determine detrimental effects, then it should have thresholds identified.

Analyzing Data — Good data analysis depends on having followed the preceding steps in the assessment design. A valuable analytical tool is scenario planning, which tries to answer a series of "what if" questions to explore the degree of difference among a spectrum of goals. Scenarios should develop future possibilities but not necessarily future alternatives. There is always a chance that scenarios formulated directly as alternatives will reduce decision space.

SELECTED READING

Allen, T.F.H., R.V. O'Neill, and T.W. Hoekstra. 1984. Interlevel Relations in Ecological Research and Management: Some Working Principles from Hierarchy Theory. USDA Forest Service General Technical Report RM-110. Rocky Mountain Forest and Range Experiment Station, Fort Collins, CO.

Botkin, D.B. 1990. *Discordant Harmonies: a New Ecology for the Twenty-first Century.* Oxford University Press, New York.

Bourgeron, P.S. and M.E. Jensen. 1994. An Overview of Ecological Principles for Ecosystem Management. In: Jensen, M.E. and P.S. Bourgeron (eds.), Eastside Ecosystem Health Assessment — Volume II: Ecosystem Management: Principles and Applications. USDA Forest Service General Technical Report PNW-GTR-318. Pacific Northwest Research Station, Portland, OR.

Burkett, E. L., and G. Lessard. 1996. Lessons Learned Workshop: Policy, Process, and Purpose for Conducting Ecoregion Assessments. USDA Forest Service, Washington, DC.

Haynes, R.W., R.T. Graham, and T.M. Quigley (eds.). 1996. A Framework for Ecosystem Management in the Interior Columbia Basin. USDA Forest Service General Technical Report PNW-GTR-374. Pacific Northwest Research Station, Portland, OR.

Hirvonen, H. 1992. The Development of Regional Scale Ecological Indicators: A Canadian Approach. In: D.H. McKenzie, E. Hyatt, and J.V. McDonald (eds.), *Ecological Indicators.* Elsevier Science Publishers, New York, NY.

Holling, C.S. 1978. *Adaptive Environmental Assessment and Management.* John Wiley and Sons, New York, NY.

Lackey, R.T. 1996. Seven pillars of ecosystem management. *Landscape and Urban Planning.* January: 1–17.

Lee, K. 1993. *Compass and Gyroscope. Integrating Science and Politics for the Environment.* Island Press, Washington, DC.

Quigley, T.M., R.T. Graham, and R.W. Haynes, eds. 1996. Integrated Scientific Assessment for Ecosystem Management of the Interior Columbia Basin and Portions of the Klammath and Great Basins. USDA Forest Service General Technical Report PNW-GTR-382. Pacific Northwest Research Station, Portland, OR.

Southern Appalachian Man and the Biosphere (SAMAB). 1996. The Southern Appalachian Assessment Summary Report: Reports 1–5. USDA Forest Service Southern Region, Atlanta, GA.

Sierra Nevada Ecosystem Project (SNEP). 1996. Status of the Sierra Nevada: Volume 1 Assessment Summaries and Management Strategies. Wildland Resources Center Report No. 36. University of California, Davis, CA.

Thorton, K.W., G.E. Saul, and E.D. Hyatt. 1994. *Environmental Monitoring and Assessment Program: Assessment Framework.* U.S. Environmental Protection Agency, Research Triangle Park, NC.

Walters, C.J., and C.S. Holling. 1990. Large-scale management experiments and learning by doing. *Ecology* 71(6): 2060–2068.

Monitoring and Evaluation

Why Monitoring and Evaluation Are Important To Ecological Stewardship

Monitoring and evaluation get a rhetorical nod from practically everyone in natural resource management. In practice, monitoring and evaluation have been frequently shunned, ignored, and underfunded by resource managers and policy makers alike. Ecological stewardship is nevertheless utterly dependent on good monitoring and evaluation systems. Complex systems, incomplete data, and uncertainties about the effects of management actions mean that monitoring and evaluation are not luxuries but essential components for any ecological approach to managing natural resources (Box 1). Greater demand for resources, increased public involvement in management, and growing concerns about sustainability of species and ecosystems, means that guessing will no longer do (Fig. 1). Society wants to better understand its land and water resources and how they are changing over time. To achieve this understanding requires monitoring and evaluation.

This summary was drafted by Robert Szaro with contributions from David Maddox, Timothy Tolle, and Mary McBurney. It is based on the following chapters in Ecological Stewardship: A Common Reference for Ecosystem Management, Vol. III:

Tolle, T., D.S. Powell, R. Breckenridge, L. Cone, R. Keller, J. Kershner, K.S. Smith, G.J. White, and G.L. Williams. "Managing the Monitoring and Evaluation Processes."

Maddox, D., K. Poiani, and R. Unnasch. "Evaluating Management Success: Using Ecological Models to Ask the Right Monitoring Questions."

KEYWORDS: monitoring, evaluation, ecological model, indicators, resource trends

Fig. 1. The black-footed ferret, an endangered species, has been the focus of intensive monitoring efforts. (Photo courtesy of US Fish and Wildlife Service.)

Organizations monitor ecological systems for a variety of reasons: to track trends in resources, evaluate management actions, and provide timely warning of undesirable conditions. They also monitor to generate timely and helpful answers to questions about the management and stewardship of their biological resources.

Natural resource managers are generally charged with the mid- to long-term stewardship of land and natural resources. This stewardship involves sustaining such landscape and ecosystem features as species populations, landscape mosaics, and commodity production. Stewardship requires management actions to maintain or correct ecosystem trajectories, and periodically assessing and monitoring in a continuous cycle. Thus, timely monitoring and evaluation occurs with a periodicity that allows the productive correction of management actions. Helpful monitoring provides information that is explicit and focused on particular issues and problems, answering

---

**Box 1
Definition of Monitoring**

- *Monitoring* is the systematic observation of parameters related to a specific problem, designed to provide information on the characteristics of the problems and their changes with time (Spellerberg 1991).
- *Ecological monitoring* is the acquisition of information to assess the status and trend of the structure and functioning of biological populations and communities, and their habitat, and larger-scale ecosystems (i.e., landscapes) over time, for the purpose of assessing and directing management activities.

specific questions about the status of resources and the effectiveness of management. It facilitates change detection to determine whether we are diverging from conservation and resource management goals.

In a democracy, monitoring can play a vital role by evaluating how the government spends and manages public resources. A sound, effective, and participatory monitoring program promotes public trust and support.

KEY FINDINGS

✓ **Monitoring programs must focus on indicators that shed the most discriminating light on critical processes and management actions.** Ecosystems, which tend to be complicated and span large areas, include a diversity of species, communities, and functions. We cannot measure everything, however, so we must keep track of those selected things or processes most indicative of the changes we seek to encourage or to avoid.

✓ **Poor data can lead to incorrect choices and expensive management mistakes, or misleading assessments of ecosystem threats.** Designing an effective monitoring program within limited budgets is a challenge, but the stakes are real. In these situations, statistical interpretation mistakes are possible. It is important to be candid about these mistakes and minimize the ones that would be the most catastrophic.

✓ **Developing an ecological model and monitoring program requires a process of intuitive and inductive thought.** The first stages of this process involve acquiring a preliminary understanding of key components and processes and the spatial and temporal scales at which they operate. For most ecosystems, researchers have to rely on incomplete and patchy knowledge. Their sources of information include previous research, particularly historical patterns and other management experience, comparison of similar ecosystems, and biological experience and intuition.

✓ **Monitoring, management, and conceptual understanding, embodied in the ecological model, are one collective iterative process.** One should develop an ecological model, and its rendering of current ecological understanding, before starting plans for management and monitoring. Once management and monitoring have begun, the ecological model can become the "table" around which to discuss management results in the light of monitoring data. Data interpreted in terms of the model serve as the raw material of adaptive management by inspiring modifications to management actions, and even to the model itself.

✓ **An ecological model suggests which entities and processes are most critical for monitoring.** The next step is to craft the best indicators. When indicators follow from components of the ecological model, then analysts can interpret data in terms of ecosystem process and management actions. New data can also provoke the analyst to modify the model itself. This interplay is the essence of adaptive management.

✓ **Constructing ecological models and monitoring plans should not occur separately from management and planning.** Ecological models express a progression of scientific thought that starts with determining key ecological components and ends with a summary of the causal ordering and relationships among them. Ecological models are maps or flowcharts that help

---

Box 2
Three Roles of Ecological Modeling in Monitoring Programs

1. A model summarizes the most important ecosystem descriptors, spatial, and temporal scales of biological processes and current and potential threats to the system. That is, the model explicitly summarizes the *current* biological understanding of the system. The model (a) does not represent the truth; (b) is not final or unmodifiable; and (c) is not expected to be complete or include the entire ecosystem. It is a flexible framework that should evolve as understanding of the ecosystem increases.

2. A model plays an important role in determining indicators for monitoring. Because the model is a statement of important biological processes, it therefore identifies aspects of the ecosystem that should be measured. If the model is a good reflection of current understanding, but the measurement indicators cannot be seen in the model, then the measurements do not have much to do with the ecosystem.

3. A model is an invaluable tool to help interpret monitoring results and explore alternative courses of management. An explicitly stated model is a summary of current understanding of and assumptions about the ecosystem. As such, it can motivate and organize discussion and serve as a "memory" of the ideas that inspired the management and monitoring plan.

---

navigate management direction and interpret results, guiding monitoring programs in three general ways (Box 2).

PRACTICAL CONSIDERATIONS

Box 3
Steps for Constructing an Ecological Model

1. Gather and assemble relevant data, information, and knowledge on system components and whole-system processes.
2. Decide on structure of model.
3. List all important states, transitions, entities, and threats.
4. Illustrate known and record unknown relationships among system states.
5. Discuss draft and revise as needed.
6. Send out model for review.
7. Update and improve model as new information becomes available.

Developing a monitoring program involves several actions. These include: (1) defining monitoring objectives early and clearly; (2) understanding pattern and process; (3) deciding essential properties; (4) arranging essential properties into an ecological model (Box 3); (5) determining the best indicators for specific aspects of the model; (6) choosing a sample design, tied to the objectives and required level of precision, and; (7) tying evaluation to monitoring objectives and decisionmaking. A monitoring program should not be implemented in isolation from other activities. Modeling the ecosystem, monitoring the status of the ecosystem, and evaluating management actions are all part of one integrated and iterative process. One should design and implement monitoring activities in conjunction with management actions or developments that are likely to affect the ecosystem.

How inclusive should ecosystem monitoring be? Monitoring should be limited to essential features of the ecological system, which include the central formative and preservative processes illustrated by the model. A fundamental tension always exists between creating comprehensive models of ecosystem behavior with fully estimated parameters and identifying a single key feature to monitor, which can serve as a canary in the ecosystem's coal mine. At its extreme, this dichotomy is the choice between broad (but shallow) and detailed (but hopelessly narrow). An expensive model that is too complicated to estimate or even specify represents the former, while the latter provides information on only a fraction of the total ecosystem. We require indicators that are focused, but still represent the breadth of the system, which typically means we need multiple indicators. A miner could monitor his coal mine with a canary because he referred to a clear conceptual model that identified problems (the bird is dead), diagnosed causes (the air is bad), and prescribed actions (get out of the mine). In complicated ecosystem monitoring, we should monitor multiple canaries and use a well-crafted ecological model to justify their selection.

Good indicators have three general scientific functions. These include assessment of ecosystem status, prediction of future problems (i.e., "early warning"), and diagnosis (Box 4): No single indicator will fill all three roles; therefore, it is best to use multiple indicators that are complementary and together best serve the evaluation of management. Indicators at lower levels of biological organization will tend to be most diagnostic, although they have certain limitations.

An ecological model can directly suggest the entities to monitor. It then remains to determine the best indicators for these entities. When indicators follow from components of the ecological model, we can clearly interpret data in terms of ecosystem process and management actions.

Developing clear, quantitative thresholds or magnitudes of significant change is critical (Box 5). Otherwise, changes observed in indicator values are incomprehensible and one cannot evaluate management. Setting explicit thresholds and magnitudes of significant change greatly facilitates the sampling and statistical design of monitoring programs.

The design of a monitoring program should build upon the goals and objectives of the manager. In this situation, the designer of the program fits the appropriate science and technology to the questions being asked. A scientifically sound sampling design is required to provide the necessary information. A poorly designed monitoring program can be worse than none at all. Before planning monitoring programs, one should address the following questions: (1) What do I need to know? (2) What will I do with the information obtained? (3) How much change is important to detect? (4) How confident do I need to be in my conclusions?

Evaluation is the process of converting monitoring data into information, and then into knowledge. Evaluation is a value-added process that provides managers with what they need to make sound decisions. There are two critical points about evaluation. First, monitoring studies must be designed to evaluate something specific. Unfocused data collection that has no clear evaluative purpose is a waste of resources, because no benchmarks exist for success, failure, or most importantly, conclusion. Second, while evaluation occurs after the actual data are collected and analyzed, the evaluation process should be determined at the outset when objectives are set. Evaluation is more likely to be successful if the

---

**Box 4
Three Classes of Indicators Used in Monitoring**

1. Assessment indicators allow simple temporal tracking of ecosystem character or comparisons of observed ecosystem attributes to expected or hoped for values.
2. Predictive indicators give warning of ecosystem stress.
3. Diagnostic indicators enrich interpretation of the causes of ecosystem changes.

**Box 5
Three Reasons to Determine Quantitative Ecosystem Monitoring Standards**

1. They are easy to communicate, and thus are better for inspiring debate and discussion about their appropriateness.
2. They provide a clear benchmark for evaluation, either in the form of "success" (one has restored ecosystem values to some predetermined level) or concern (the ecosystem has changed by a significant degree).
3. They routinize many difficult sampling and statistical issues, especially with regard to the sampling intensity required to adequately measure the biological changes and patterns expected.

Box 6
Items To Be Considered When Identifying The Purpose(s) and Objective(s)
of Monitoring

- Monitoring objectives must be realistic and attainable (i.e., cost effective and able to be obtained in the time frame for which the information is needed).
- Focus the objectives on solving a definable problem or issue. One can't measure everything.
- Ask what questions interest the decisionmaker and the general public. Ask the users of the data to identify the monitoring purposes.
- Develop the objectives in such a way that decisions can be based on monitoring results. Objectives can be phrased as questions to be answered during evaluation.
- Determine the scope of the monitoring as part of tightly defining the purpose. Project monitoring will usually focus on ripe issues. Broader landscape-scale monitoring often focuses on stewardship and sustainability over longer temporal scales. Examples of the latter include determining the status and trends of ecosystem integrity and forest health.
- Evaluate existing data to see if they can answer the monitoring questions. If not, develop a proposed evaluation method.
- If the monitoring questions could not be analyzed or evaluated from the data that would be collected, reconsider the objective as well as the method.

---

criteria of evaluation are worked out in advance as part of an overall concept of management and monitoring. Criteria for evaluation involve decisions about what variables to measure (i.e., study design) and what results constitute management success or failure. In other words, after you have collected the data, how will you recognize success and failure of the management?

Institutionalizing monitoring and evaluation into land-management practices is key. To be effective, an organization must endorse monitoring both from the top down (i.e., the organization's leaders must support and approve the work) and from the bottom up (i.e., the work must be practical and pertinent). The monitoring process should include outside experts and public involvement. Institutionalizing effective monitoring practices requires consistency in handling and long-term maintenance and preservation of data, information, and samples.

To be meaningful and useful, monitoring objectives should: (1) state fully what the work intends to accomplish, and; (2) specify a recognizable end point to determine the progress or attainment. Thus, it is important that people clearly state monitoring objectives, using unambiguous wording to ensure that there are no questions about what they are measuring or monitoring. Good monitoring objective statements, which are generally stated in quantitative terms, provide a yardstick against which others can evaluate the progress made (Box 6).

Evaluate existing data before rushing out to collect more. After setting the monitoring and evaluation objectives and identifying data needs, the next step is *not* to begin collecting new data. This costly fault has plagued agencies for years. The first implementation step is to search for useful existing data. This search may identify previous successes and failures, determine local variability for setting the proper sample size, or identify alternative measures from the literature. If the data do not exist and if no one else is collecting similar data elsewhere, then it is appropriate to design a monitoring system.

The many types of monitoring can be categorized in numerous ways, including detection, evaluation, and compliance. Monitoring can also be categorized by ecosystem or resource emphasis (e.g., fish, wildlife, vegetation), scale, process, or in other ways. The objectives differ for each of these types of monitoring, but three key questions cover all of them: (1) Did we perform the management actions we said we would (implement)? (2) How well are we meeting management goals (effectiveness monitoring)? and (3) Are the key assumptions valid (validation monitoring)?

Baseline monitoring can provide information for understanding unexpected ecological events of any given place. For example, when the Exxon Valdez incident occurred, very little baseline information was known, which made it difficult to assess damage, prioritize restoration and cleanup.

Quality control is an essential component of monitoring and evaluation. Quality control is the routine application of prescribed field and office procedures to reduce random and systematic errors and ensure that data are generated within known and acceptable performance limits. Monitoring data must be of sufficient quality to adequately meet the objectives established at the beginning of the process. If they are not, they are worthless.

The flow of information must be managed from the time it is recorded in the file or on the map until the summary statistics are examined. This statement seems obvious, but the step is often neglected. Some even say the lion's share of the monitoring cost should be in this information management step, but this is where people usually spend the least money and time. The evaluation should use appropriate analytical tools, be statistically matched to the stated objective(s), and be simply communicated.

Adequate documentation of monitoring data and evaluation results will significantly contribute to institutionalizing this important work. Consistency in data handling and preservation of data, information, and in some cases samples, are essential to institutionalize effective monitoring practices (Fig. 2). Plans must be implemented to ensure long-term maintenance and preservation of monitoring data. The archival method for monitoring information should not be manager- or agency-specific but should maximize its potential for different groups to share or later compare the information with more recent information. In many

Fig. 2. Monitoring often requires extensive sampling. The data must be consistently collected and carefully compiled to ensure the maximum use of the results. (Photo courtesy of USDA Forest Service.)

cases, it may be necessary to preserve maps and permanent ground markers. Funds should be earmarked for the proper documentation of monitoring information.

Results of monitoring and evaluation should be clearly presented. Easily understood forms of communication may involve tables, maps, graphs, charts, reports, models, performance indicators, or presentations, such as slide shows or videos. All the hard work and expense of monitoring and evaluation are for nothing if the intended audience, which may include researchers, policy-makers (including the public), and managers, does not use the results for their intended purpose.

The credibility of monitoring efforts is essential from two points of view. First, the public needs to understand the importance of monitoring and how public use affects their lands and resources. Second, the public needs to understand how management decisions and land-management policy are often based on the outcome of monitoring efforts. Managers of public lands will achieve credibility to the extent that they keep their partners and publics involved in all phases of the monitoring and evaluation process. Anyone can produce data and try to impress people with them. Managers of public lands have a responsibility to provide the citizens of the United States with the best information possible. The public will view as most credible the monitoring data that are collected using the best scientific knowledge, have known precision, are of highest quality, and are as objective as possible. Proper monitoring and evaluation can help public land managers regain lost public trust.

SELECTED READING

Fairweather, P.G. 1991. Statistical power and design requirements for environmental monitoring. *Australian Journal of Marine Freshwater Research* 42: 555–567.

Green, R.H. 1979. *Sampling Design and Statistical Methods for Environmental Biologists.* John Wiley & Sons, New York.

Hayek, L.C. and M.A. Buzas. 1997. *Surveying Natural Populations.* Columbia University Press, New York.

Holling, C.S. 1978. *Adaptive Environmental Assessment and Management.* John Wiley & Sons, New York.

Keddy, P.A. and C.G. Drummond. 1996. Ecological properties for the evaluation, management, and restoration of temperate deciduous forest ecosystems. *Ecological Applications* 6: 748–762.

Kenkel, N.C., P. Juhasy-Nagy, and J. Podani. 1989. On sampling procedures in population and community ecology. *Vegetatio* 83: 195–207.

Spellerberg, I.F. 1991. *Monitoring Ecological Change.* Cambridge University Press, Cambridge.

Suter, G.W. 1993. A critique of ecosystem health concepts and indexes. *Environmental Toxicology and Chemistry* 12: 1533–1539.

Zar, J.H. 1984. *Biostatistical Analysis.* Prentice-Hall, Englewood Cliffs, NJ.

Data Collection, Management, and Inventory

Why Data Collection, Management, And Inventory Are Important To Ecological Stewardship

Today's ecosystem management planners face a set of management complexities not encountered by their predecessors. In the not too distant past, natural resource managers surveyed the terrain and its resources from atop a horse and recorded their observations in a notebook they secured in their shirt pocket. They relied on the notes they took and the maps they drew to manage timber, wildlife, and minerals as a select and independent set of resources. The advent of pickup trucks, computers, satellites, environmental laws, and a more diverse and informed public has introduced a complex set of management demands, endless volumes of data, and new ways to collect, organize, and use information. Data — their collection, quality, integration, storage, and access — have become critical to achieving successful results in natural resources management. Ecosystem approaches, in particular, depend on the sound collection and management of information because they simultaneously deal with multiple resources, and often seek to address environmental, social, and economic objectives.

KEYWORDS: Data, inventory, monitoring, information technologies, knowledge

This summary was drafted by Robert Szaro with contributions from Allen Cooperrider, Cynthia S. Correll, Joe Lint, and Andrew Malk. It is based on the following chapters in Ecological Stewardship: A Common Reference for Ecosystem Management, Vol. III:

Cooperrider, A., L. Fox, III, R. Garrett, and T. Hobbs. "Data Collection, Management, and Inventory."

Correll, C.S., C.A. Askren, R. Holmes, H.M. Lachowsky, G.C. Panos, and W.B. Smith. "Managing Information for Ecological Stewardship."

Collecting data in and of itself is not enough for making informed decisions. Intermediate steps are required to transform data into information and to convert information into knowledge. Informed decisions result from knowledge, not data. The data–information–knowledge linkage is important not only for today's managers to make informed decisions, but it also provides the baseline for decision-makers of tomorrow. With it, they can open notebooks from the past to better understand what others have done before them and to assess the progress of ecosystem management.

KEY FINDINGS

Ecosystem management demands a long term investment in the collection and preservation of data. The data collected today become the data of the past, and grow in value and relevance with each passing day — provided they are maintained. Data management is essential to ensure the return on investment in data collection, and to prevent a bankruptcy of the knowledge needed to be successful in ecosystem management.

✓ **A recurring problem in natural resource management is that data collection is used as an excuse for inaction**. Science, research, studies, inventory, and monitoring have often been seen by the public and resource managers as code words for delay. Politicians, for example, have consistently used the need for more information as an excuse for not dealing with well-documented problems of the ecological effects of acid rain. One can find the same pattern with global warming, depletion of oceanic fisheries, and more local issues. Committing resources to data collection provides policy makers with the cover of "good intentions" for taking no action. However, inaction is also a management decision, which often causes increased risk of ecological destruction.

✓ **Another problem is that the vast amount of data collected is seldom used effectively**. On a broad scale, the current problems of the Pacific Northwest forests (endangerment of spotted owl, marbled murrelet; collapse of salmon and other anadromous fisheries; loss of old growth forest; declining forest health; polarization of interests; decline of forest industry) developed in spite of extensive research and studies on all of these issues. Again, numerous examples of more specific, local issues could be cited. Collecting, sharing, and using resource data often consume over half the cost of ecological stewardship projects. It is important to manage that investment carefully.

✓ **There are also numerous cases in which too little data are collected (or not collected to reliable standards) even in the face of known needs for data or information** (Box 1). In such situations, managers typically rely upon ideology rather than information to make decisions.

Factors limiting effective use of data

Several factors often limit the effective use of data. The most prominent of these factors include:

✓ **Lack of commitment to the use of science**. In natural resource management, as in other functions of our modern society, there is often a lack of commitment to the use of data/information or science to resolve

---

Box 1
Common Problems Facing Data Management Efforts

- Data collection (science, research, studies, inventory, monitoring) is used as an excuse for inaction.
- The vast amount of data collected are never used effectively.
- The wrong data are collected.
- Data are collected at the wrong temporal or geographic scale.
- Fragmented disciplinary approaches to ecosystems have led to different definitions and names for the same ecosystem components, science, and so on.
- The half-life of both scientific and management data is short.
- "Information" (summarized or synthesized data) is stored/archived whereas the raw data is not. The most obvious and useful information (such as species distributions) typically has not been compiled into range maps.
- Data collected at one geographic scale is used to draw conclusions at another scale.
- The method of communicating scientific information is outdated and in need of major renovation.
- Managers and archivers of data need to be aware of, and anticipate if possible, the potential legal pitfalls resulting from the way they collect and keep data.

---

conflicts and make decisions. If policy and management decisions are being based largely upon ideology or politics, then the best data and the best science are unlikely to help.

✓ **Inadequate "framing of the question."** It is well known among those in politics that control of the agenda is the surest way to ensure that one's programs and policies are supported. Thus the person who can "frame the question" is in control of the debate. Scientists have often allowed the politicians and managers to frame the question, when they should have insisted that scientists also be involved.

✓ **Unclear objectives.** As in most endeavors in life, unclear objectives can lead you almost anywhere. The problem begins with management objectives; if these are unclear, then developing a supportive science program is unlikely to be successful.

✓ **Information transfer.** The communication of information from one subculture (scientists, managers, citizens, etc.) to another is perhaps the weakest link in the whole process. Each subculture has its own set of jargon and preferred methods of communication. And although the importance of cross cultural communication is paramount, there is virtually no reward system or encouragement for effective cross-cultural communication (Fig. 1).

✓ **Scientific reward system.** Science is founded on a reward system that is peer-centered rather than client-centered. This is the root cause of the irrelevance of much scientific work

✓ **Scale of scientific work:** The scale of scientific work is not aligned with the scale of political decisions. As pointed out earlier, science is often focused on small geographic areas or short time-frames. However, even large scale science may be poorly matched with the spatial scales at which most decisions are made.

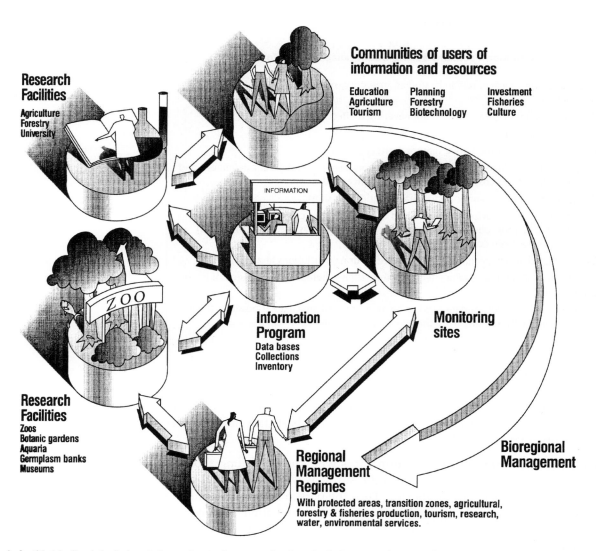

Fig. 1. In this idealized depiction, information is shown to circulate freely between distinct information communities and cultures.

New technologies

New technologies are rapidly transforming the way we collect and store data. These technologies offer enormous benefits, but they also can create problems that resource managers should be aware of.

✓ **Data technologies involve trade-offs**. While technologies may help with some people and some tasks, they also cause problems. One of the purported strengths of many data management technologies is that they alleviate the need to go to the field. But this is a mixed blessing. The field can also be a good place for discussions between and among scientists, managers, and citizens. Direct observation of the world needs to be the starting point for data management; data management technologies are merely designed to organize and archive our observations. The ultimate strength of any such technology will depend on the base of observations upon which it operates.

✓ **Remote sensing**. As we move into the age of remote sensing, we tend to become removed from the very subjects of interest — the forests, rangelands, and aquatic ecosystems we are trying to study, understand, and

Fig. 2. Each new Landsat satellite adds exciting capabilities to remote sensing. This should not, however, distract us from the underlying goal of interacting directly with the very systems we are trying to understand and manage — the forests, rangelands, and aquatic ecosystems. (Photo courtesy of NASA.)

manage (Fig. 2). We need to remind ourselves that the purpose of such remote observation is to understand what we see on the ground. Remote sensing, as with other technologies, is not an end in itself.

✓ **Geographical Information Systems (GIS)**. When GIS data layers are compiled seamlessly throughout an entire ecologically significant region, several benefits accrue (Box 2).

✓ **Internet**. Although the "Net" presents a powerful tool with endless possibilities for transfer of data and information, it is not without its limitations: (1) it is not accessible to everyone, (2) as a means of transmitting scientific information, it bypasses the peer review process, and (3) it does not substitute for direct communication and observation.

Information Planning

Good information management doesn't just happen; it requires advance planning.

Box 2
Benefits of Seamlessly Integrated GIS Data Layers

1. Any cooperating agency can view geographic features throughout the entire hydrobasin and construct thematic maps of ecosystem condition for basins and drainage within basins.
2. Baseline data can be archived for distribution to cooperators.
3. Geographically-referenced ecosystem information will allow cooperating agencies to analyze the spatial statistics of environmental variables such as area by habitat type, interspersion, habitat heterogeneity, etc.
4. Referenced, multi-temporal data sets addressing environmental variables will allow cooperating agencies to conduct their own analysis of data to monitor ecosystem condition for the entire geographic domain of interest.
5. Management alternatives, including the geographic relationships of potential outcomes, can be evaluated in advance.

Box 3
Life Cycle Management Phases

- *Initiation* — The purpose of the initiation phase is to have a clear understanding of the objectives, benefits, scope, personnel, authorities, reporting relationships, deliverables, and resources needed to carry the project through its entire life cycle.

- *Development* — Development usually involves: (1) analysis, (2) design, and (3) buy or build the solution.

- *Implementation* — The implementation phase, sometimes also called "transition" or "rollout", is the most critical part of the whole project.

- *Operations and Maintenance* — This final phase is to support users in their efforts to efficiently access, use, and manage the information they need for their work.

✓ **Different approaches to information planning are appropriate for different ecosystem management endeavors.** For efforts such as fire recovery projects or watershed analyses, a common approach to information planning is the information needs assessment (INA). INAs most often focus on analysis of all the spatial data needs of a specific project. They have breadth but not depth. A different approach is needed when a group of agencies, an agency, or a unit of an agency, embarks on building a specific long term subject-matter database that will serve the needs of multiple future projects. Examples of such databases are a vegetation or an ecological classification and inventory database. Flexibility and long term support are key considerations. Planning needs for this type of effort are better met by a structured life-cycle planning process.

✓ **Life-cycle planning.** Managers have little confidence that major long-term systems development will be done on time and within budget. History backs up their pessimism. Developers have a history of promising to do too much too quickly for too little. Developing long-term, flexible, multi-use data stores and systems takes major investment and time. And these are the kind of systems highly valuable to ecosystem management, where data are used by multiple functions, and are needed for monitoring change and trend over time, and to answer questions both known and yet to surface. An often overlooked fact is that development cost has historically been less than 20 percent of the total life-cycle cost of a system. Support and maintenance have typically grabbed 80 percent of life-cycle cost. A variety of information planning techniques, standards, and tools that fall under the general heading of "life-cycle planning" can be used to better estimate costs and time lines, improve information products, and lower support and maintenance costs. At a minimum, a manager needs to have a feel for the structured steps (or phases) of life-cycle management, and the roles that need to be implemented. Following these steps or phases and establishing these roles will help control typical problems associated with large projects, including failure to keep users involved, lack of accountability, hidden costs, lack of completion, over commitment of resources, failure to maintain systems, and frequent changes to requirements (Box 3).

PRACTICAL CONSIDERATIONS

Data collection and inventory. Part of the overall mission of all public land management agencies is to collect, process, and analyze data to improve understanding and management of our natural resources. As part of this mission, each agency is generally charged to: (l) meet data user needs for resource information, (2) cooperate with interested individuals and organizations in planning resource inventories, (3) interact with international organizations in coordinating global assessments, and (4) provide for training and scientific information exchange within each agency, and in a larger sense, within the global community. Inventories are the baseline for monitoring by providing a comprehensive review of the status and trends of our diverse ecosystems, their use, and their health.

Implementation. Selecting appropriate sampling approaches, methods, and tools will depend primarily on the initiative of the dominant users of the data to be collected, and their willingness to view the larger 'information landscape' in the decision process. Accepting a functional approach to fulfilling a need for

Table 1. Hierarchical Relations Between Assessment Scales, Types, and Various Ecosystem Delineations

Assessment Scale	Biophysical Environment		Existing Conditions	
	Terrestrial Units	Aquatic Units	Vegetation Units	Social Units
Global	Domain	Zoogeographic Region	Class	Continent
Continental	Division	Zoogeographic Sub-region	Subclass	Nation
Regional	Province	River Basin	Group	State
Sub-Region	Section/Subsection	Sub-basin	Formation	County
Landscape	Landtype, Association	Watershed	Series	Community
Land Unit	Landtype, Landtype Phase	Valley Section, Stream Reach	Association	Neighborhood
Site	Ecological Site	Channel Unit	Group	Household

inventory information is neither good nor bad if the investigator has evaluated broader alternative approaches and found them less efficient from either the information or financial point of view. This will be especially true of one time event or limited scope issues, or when temporal issues prevail.

Sampling approaches. Generally speaking, approaches to data collection have less to do with a particular discipline and more to do with the physical parameters of source, content, and scale of the data to be collected. Analysis applied to a scale for which the data are not matched or inappropriately integrated can be disastrous. Data collection approaches can be seen as a continuum ranging from purely functional with a single variable of interest at a single point in time, to fully integrated with multiple variables and scales over multiple time frames. This continuum can be arbitrarily broken into three discrete categories for simplification: (1) functional, (2) linked-functional, and (3) integrated/interdisciplinary. Approaches must also seek to standardize terminology by first identifying output products, then defining what variables are needed to produce those products. Existing terms and information should be used whenever possible to develop a framework for organization, compilation, and presentation. One way to ensure that the most appropriate approach is used is to view the opportunities for integration from a more global perspective. Table 1 provides a simplified representation of how scale and objective can be viewed.

Sampling methods and tools. Closely allied with the selection of a sampling approach is the selection of methods and tools for gathering the data. There are four basic sampling methods that singly or in combination can be used to gather data (Box 4).

Identify data requirements based upon a clear understanding of the need. Apply some form of "information needs assessment" or "decision science" to determine what data themes are required and at what scale, accuracy, and precision (Box 5). Reuse data to the extent possible. Before collecting new data, search existing

---

Box 4
Basic Sampling Methods

1. *Census* — a complete enumeration of the population of interest. If economical, this method is highly desirable because it minimizes the risk that you do not have a representative view of the population of interest. This approach is most useful when the population is small or the data so critical that the expense is justified.
2. *Mapping* — surface features of a landscape, benthic(underwater) terrain, or object using remote sensing methods include photography, videography, satellite imagery, radar, sonar, laser imaging. The photography and videography may be from either aerial or ground perspective.
3. *Non statistical sample* — subjective, often cost-effective precursor to statistical sampling. Cunia (1982) lists four situations in which this form of sampling may be preferred: (1) variations between elements of the population are large and sampling is expensive; (2) the needs for information about a population are immediate and a decision must be made before a well-executed statistical sample can be executed; (3) funding is short or unavailable and the only alternative is to use existing information and extrapolate to the population of interest; and (4) approximate knowledge of some of the population parameters are needed to design an efficient statistical sample.
4. *Statistical sample* — the collection of data, representative of the population of interest, in a scientifically acceptable manner. The sample is generally probabilistic in nature with every unit of the population having a positive probability of selection and these probabilities are known and are independent of the person taking the sample.

---

local, regional, and other sources. Assess available data standards and apply those most appropriate for the job at hand.

Keep it simple. Collecting data from many sources or attempting to use the latest technology may sound attractive, but it can also be costly in terms of time and complexity. Ensure a good mix of resource and technical knowledge, skills, and abilities on any data management activity. Knowledge of networking helps bring people and data together across boundaries. GIS applications and remotely-sensed data are critical to most resource management activities today.

There are no "right" answers in data management. Only the possibility of increasingly "better" answers. The notion of adaptive management applies to data management, just as it does to ecosystems management.

Data standards

As the forms and types of data for natural resources management have proliferated, the need for consistent standards has become obvious. Several efforts to establish data standards are worth noting.

The National Spatial Data Infrastructure (NSDI). The NSDI has been established in response to the need for people and organizations to be able to better share in the collection, use, and maintenance of spatial data. Geographic data have always been important to natural resource managers and scientists, but with the rapid expansion and use of geospatial data technologies, there are more producers and users than ever, and much more involvement with data through electronic networking and shared/collaborative decision making. Along with wider use of these new technologies has come a growing recognition that geospatial data issues must be addressed to reduce total costs and to make cooperative use and sharing of data a reality. The NSDI is the way

---

Box 5
Considerations for collecting inventory data when functional data or approaches exist:

1. What is the scale and quality of existing data?
2. Does the study require acquisition of new data?
3. Is there a narrow or broad need for the data?
4. What are the periodicity and scale issues related to the data?
5. Are the data of use to other disciplines?

in which the United States is organizing and coordinating its geospatial data activities. The NSDI (established by Executive Order 12906 on April 11, 1994) directs Federal agencies to provide leadership in its development and partnership with State, local, and tribal governments, academia, the private sector, professional societies, and others.

The Federal Geographic Data Committee (FGDC). The FGDC is an interagency committee that promotes coordinated development, use, sharing, and dissemination of geospatial data on a national basis through the NSDI. The FGDC provides the Federal leadership in working with all other partners, and in initiating and coordinating work activities crucial to the implementation of the NSDI. Key areas include the framework, a base upon which to collect, register, and integrate information; the standards, both thematic and geospatial; and partnerships.

SELECTED READING

Antenucci, J.C. 1991. *Geographic Information Systems: A Guide to the Technology.* Van Nostrand Reinhold, NY.

Cunia, T. 1982. The needs and basis of sampling. In: *In-Place Inventories: Principles and Practices. Proceedings of a National Workshop, Orono, ME, 1981.* pp. 315–320. Society of American Foresters, Bethesda, MD.

Davis, B. 1996. *Geographic Information Systems: A Visual Approach.* Onword Press, Santa Fe, NM.

Domanski, P. and P. Irvine. 1996. *A Practical Guide to Relational Database Design.* Domanski-Irvine Book Co., Herefordshire, UK.

Hutchinson, S. and L. Daniel. 1995. *Inside ArcView.* OnWord Press, Santa Fe, NM.

Jensen, J.R. 1996. *Introductory Digital Image Processing.* Second Edition, Prentice Hall, New Jersey.

Morain, S. and S.L. Baros (eds.). 1996. *Raster Imagery in Geographic Information Systems.* OnWord Press, Santa Fe, NM.

Robert, Karl-Henrik. 1993. Educating a nation: the natural step. *Context* 28: 10–15.

Sample, V.A., ed. 1994. *Remote Sensing and GIS in Ecosystem Management.* Island Press, Covelo, CA.

Schreuder, H.T., T.G. Gregoire, and B.W. Geoffrey. 1993. *Sampling Methods for Multiresource Forest Inventory.* John Wiley & Sons, Inc., New York.

Zeiler, M. 1995. *Inside ARC/INFO.* OnWord Press, Santa Fe, NM.

Decision Support Systems

Why Decision Support Systems Are Important To Ecological Stewardship

The impacts of natural resource management decisions on ecosystems and people are not always obvious or intuitive. In large part this is because these decisions must be made in the context of short and long temporal scales, as well as fine and broad spatial scales — scales that decision makers have little personal experience with. Moreover, decisions in ecosystem stewardship typically involve complex relationships among numerous resource and human-dimension issues, making *ad-hoc* judgments more prone to error than more formalized approaches. In many cases, experts do not understand and cannot explain all the relationships that may be involved in a decision. Careful analysis and use of good information can help improve decisions. This analysis is the heart of decision support.

This summary was drafted by Robert Szaro with contributions from Keith Reynolds and Bruce Lippke. It is based on the following chapters in Ecological Stewardship: A Common Reference for Ecosystem Management, Vol. III:

Oliver, C.D. and M.J. Twery. "Decision Support Systems, Models, and Analyses."

Reynolds, K., J. Bjork, R.R. Hershey, D. Schmoldt, J. Payne, S. King, L. DeCola, M. Twery, and P. Cunningham. "Decision Support for Ecosystem Management."

KEYWORDS: Decision theory, integrated systems, models, management science, policies

New understanding about public interests and desires has changed approaches to public land management. With ecosystem stewardship approaches, decisions now involve more diverse topics, and a larger community of interests, crossing different organizational hierarchies. Participants can significantly facilitate decision making in such a complex environment by using a variety of tools and systems in the various phases of a decision process. An important benefit from wider use of decision support methodologies is a better and broader understanding of the decision process itself.

Many researchers in such fields as defense, aerospace, management science, policy analysis, planning, and engineering have studied and developed the decision support and decision making process for many decades. From these studies, a basic, fair, open, systematic, unbiased process has evolved, which helps ensure that decisions involving complex systems lead to expected results.

KEY FINDINGS

✓ **Broad decisions made at high levels in an organization or social institution are referred to as "policies."** Decisions made at local levels (often by detailing and expanding policy decisions) are referred to as "management decisions." A continuum exists between the two levels since policies made at high levels can be interpreted differently at other levels and produce different trade-offs of values. Both levels require similar decision processes and support structures.

✓ **Decision support is the process of framing a decision as effectively as possible so that the decision maker will base the decision on the best possible knowledge of the outcomes.** Decision support, which has developed over many decades, provides the methodology and tools for framing the decision. If one understands and follows the methodology, he or she can incorporate various decision support tools to provide and analyze data, information, and knowledge. This process should improve the quality, effectiveness, and power of decisions.

✓ **Decision support methods and tools for ecosystem management are based on the combination of general systems theory and management science and the fields of decision theory and policy analysis.** The process has been emerging as an organized discipline for about 100 years under the various names of management science, policy analysis, and decision theory.

✓ **Decisions are deemed successful when the actual consequences of the decisions meet expectations — or where a different, but equally satisfying, result occurs fortuitously.** Decisions will often be successful when the perceptions of the decision makers match reality (Box 1).

✓ **The role of science in decision making is not to determine the "correct" or "best" alternative among possible actions.** Rather, scientists and policy analysts serve the decision makers and the decision making process best by attempting to remain as objective and disinterested as possible when choosing alternatives. Although complete objectivity is not possible, an honest attempt to separate the process of analysis from the choice of alternatives can help to build confidence in the reliability of the decision process.

Box 1
Two things are necessary for decision makers to make good decisions

(1) There must be some understanding of the behavior of the thing(s) to be managed. What is known and what is not known must be identified.

(2) The understanding must be accessible to decision makers in formats that allow the decision makers to understand the problem, issues, and trade-offs, weigh outcomes of alternative actions relative to a chosen set of objectives, and make an appropriate decision.

✓ **Understanding begins with precise communication**. Both analysts and decision makers need to agree on a common grouping of "modules" and flows. Similarly, people of different backgrounds often interpret words differently. Care is needed to ensure common understandings of word meanings. To avoid confusing or polarizing an issue, it is expedient to delete words that mean different things to different people and create new words.

✓ **In management and policy decisions, both objective analyses and subjective decisions are needed**. Decision makers, however, should make subjective decisions with an understanding of the probable outcomes of these decisions. Decision support systems need to be objective, both to keep the analyst from influencing the decision and to allow the systems to improve with time. Consequently, one should follow objective, analytical, systematic methods similar to those used for scientific research. Documentation and peer review of the decision support process permit decision makers and analysts to understand weaknesses in the system. They also allow the process and decisions to be improved readily as more information is gained. By making the consequences of the available alternatives explicit, decision makers' values are revealed through the alternative they choose.

✓ **Similar systematic procedures have emerged in various management and engineering science fields**. These mostly follow a general systematic procedure to ensure decisions are made with as much understanding and objectivity as possible. Sometimes several steps are combined, subdivided, or repeated as needed (Box 2).

✓ **Conceptual models, simulation models, and complex integrated systems that coordinate multiple software applications can provide decision support**. Decision support systems are not synonymous with models, but models are an important tool used in many decision support systems.

✓ **Decision makers need to understand the trade-offs and consequences of alternative courses of action**. Various optimization and priority setting tools are available to reduce complex problems into simpler components. Such analysis tools as the analytical hierarchy process, linear programs, and knowledge-based systems may help. Visual displays include matrices of alternatives versus objectives, decision trees, comparative risk profiles, and probability distributions.

✓ **Good decision support tools are easy to understand, use, and explain**. Decision support tools include interactive programs that help a decision maker understand the issues, determine goals and priorities, formulate alternative options, analyze the options, and evaluate potential consequences (Box 3).

✓ **Many decision support tools and methods can be used in combination with one another and customized to individual situations**. Although numerous decision support systems and tools exist that collectively are applicable to all phases of adaptive ecosystem management, no single system exists to provide integrated support for the full adaptive management process. Some tools and systems have straightforward and clear application to specific phases of the adaptive management process.

✓ **Following objective, systematic, analytical procedures will not guarantee that the decision making process is successful**. The process will only be successful if the expected outcomes of the decision coincide with actual outcomes and all important factors have been included. Objective

---

**Box 2
General Procedures To
Ensure Objective Analyses**

(1) Identify the problem, decision makers, their authorities, the stakeholders, and the decision making process.

(2) Define the problem, refine the objectives, and develop objectively measurable criteria to determine the degree to which each objective is reached.

(3) Develop alternative actions to achieve the objectives.

(4) Compare each alternative with the objectives.

(5) Choose a preferred alternative based on acceptable evaluation of the objectives.

(6) Implement the chosen alternative.

(7) Monitor and evaluate.

Box 3

Decision support tools can aid ecosystem management with respect to the following:

(1) Defining or developing appropriate organizational structures.

(2) Defining and identifying stakeholders.

(3) Defining and ranking objectives.

(4) Developing and describing alternative management approaches or actions.

(5) Analyzing and describing tradeoffs among alternative actions.

(6) Communicating within and across hierarchical levels in an organization

procedures and knowledgeable people can help increase the probability that the outcomes are as expected.

✓ **Various decision support tools — such as simulation models, graphics, knowledge-based systems, and even simple dichotomous keys — can help if they are used in the proper context**. The systematic organization of an assortment of tools into a computer-based decision support system can further aid the process of identifying and solving management problems.

PRACTICAL CONSIDERATIONS

Decision support tools can help to ensure public involvement. Group decision support systems are one approach to resolving conflicts. Group decision support systems use information technology to help groups of people consider uncertainty, form preferences, make judgments, and make decisions. The use of structured meeting procedures linked with such groupware systems such as AR/GIS can improve the efficiency and quality of the decision process and the likelihood that all participants will be satisfied with the outcome (Fig. 1).

Decision support tools are critical to assessment and monitoring. Knowledge-based systems, such as EMDS can be particularly useful in ecological assessment, because logic-based knowledge facilitates integrated analysis of diverse topics, including evaluation of states and processes related to biophysical, social, and economic concerns. Systems designed for assessment are equally applicable to evaluating information generated by monitoring programs that originate with implementation, thus, completing the adaptive management cycle.

Decision support tools can aid in the planning process. Decision makers are faced with the dilemma of how to allocate their scarce resources to achieve goals — such as maximizing profit from timber harvesting, maximizing wildlife habitat, minimizing travel distance, or maximizing recreational opportunities — which are often conflicting. Usually there are a large number, if not an infinite number, of possible solutions. Sometimes more than one solution exists that can provide either the same or a close objective function value. Optimization

Fig. 1. Community members review decision support tools as part of their participation in the decision process. (Photo courtesy of USDA Forest Service.)

tools, such as SPECTRUM, assist decision makers in allocating their scarce resources to achieve their management objectives. These tools include linear programming, dynamic programming, integer programming, and multiple--objective programming.

Decision support tools can facilitate implementation and monitoring. A number of project management systems exist (e.g., INFORMS, NED, and LMS) that facilitate the implementation phase of the adaptive management process, given the management context provided by a strategic plan.

Monitoring and evaluation are extremely helpful in refining the chosen alternative. They are also essential steps during implementation while preparing for the next iteration of the decision process. Monitoring and evaluation can also be used to adjust later implementation of a previous decision in the event that not all manifestations of a management decision may be implemented simultaneously. Before applying it widely, it is efficient to use early monitoring to determine if a particular action produces the expected results. Many efficient techniques for monitoring and feedback have been developed under such names as continuous quality improvement and adaptive management. Many ecosystem managers fear that monitoring and evaluating adaptive management will be cumbersome and costly. In fact, this effort can be quite inexpensive and soon becomes self-supporting because most of the information required is already collected during inventories, regeneration surveys, and other normal management practices.

For decisions regarding ecosystem management to be made fairly, efficiently, and in a timely manner, several things are needed:
- A more universal understanding of the decision making process at all levels — from the national policy to the individual area level — which allows decisions to be made and coordinated effectively;
- A central information and research base, so people can rapidly understand what information is available, what techniques are available, who can analyze a problem, and what similar procedures have been tried and have worked, and;
- A communication system to allow rapid, open communication among all levels of decision makers, stakeholders, and analysts.

Some problems that decision support technology needs to recognize as barriers to good decision making include the following:
- Decision processes often deal poorly with unanticipated outcomes, because decision makers are too focused on benefits and desired outcomes;
- People see decisions separate from organizations, because the decision making process is so diffuse;
- Many decision making tools are complex. One needs training to deal with sophisticated tools, obtain informed consultant advice and assistance, and utilize groups already well versed in specific tools;
- Many decisions are political and have more to do with power than with information, values, and perceptions;
- In group decision making, it may be difficult to develop responsibility and accountability, and stick to the decision;
- Assumptions underlying models and sensitivity of models to their parameters often are not well documented in decision support systems or specific components, and;
- All tools have the potential to be misused. Peer review or other objective processes are needed to ensure that tools are accurate and objective and used appropriately.

Some basic concepts that promote successful application of decision support technology and lead to good decision making are the following:

- Decision making is incremental. Decision makers are forming and adjusting viewpoints throughout the process;
- Decision making is dynamic. The world is changing as people are discussing decisions, so the decision making process has to be dynamic;
- Stakeholders, organizations, and agencies need to integrate and share power and responsibility. No single group has a monopoly on the necessary information, and;
- Organizations and decision makers need to provide the rationale behind decisions. With the rationale, the organization and decision makers can learn from their mistakes, because it is possible to re-examine the way they made the decision.

SELECTED READING

Adelman, L. 1992. *Evaluating Decision Support and Expert Systems.* John Wiley and Sons, New York.

Boyce, S.G. 1985. Forestry Decisions. USDA Forest Service General Technical Report SE-35.

Boyce, S.B. 1995. *Landscape Forestry.* John Wiley and Sons, New York.

Gunderson, L.H., C.S. Holling, and S.S. Light (eds.). 1995. *Barriers and Bridges to the Renewal of Ecosystems and Institutions.* Columbia University Press, New York.

Hearnshaw, H.M. and D.J. Unwin. 1994. *Visualization in Geographic Information Systems.* John Wiley & Sons, New York.

Howard, A.F. 1991. A critical look at multiple criteria decision making techniques with reference to forestry applications. *Canadian Journal of Forest Research* 21: 1649–1659.

Keeney, R.L., and H. Raiffa. 1993. *Decisions with Multiple Objectives.* Cambridge University Press, Cambridge.

Nemoto, M. 1987. *Total Quality Control for Management.* Prentice Hall, Inc., Englewood Cliffs, NJ,

Oliver, C.D., D.R. Berg, D.R. Larsen, and K.L. O'Hara. 1992. Integrating Management Tools, Ecological Knowledge, and Silviculture. In: R.J. Naiman (ed.), *Watershed Management: Balancing Sustainability and Environmental Change.* Springer-Verlag, New York.

Rudis, V.A. 1993. The Multiple Resource Inventory Decision-Making Process. In: H.G. Lund, E. Landis, and T. Atterbury (eds.), *Proceeding of the Stand Inventory Technologies: An International Multiple Resource Conference.*

Saaty, R.W., and L.G. Vargas (eds.), 1987. Special Issue Dedicated to the Theory and Applications of the Analytic Hierarchy Process. *Mathematical Modelling* 9.

Sampson, R.N., and D.L. Adams (eds.), 1994. *Assessing Forest Ecosystem Health in the Inland West.* The Haworth Press, Inc., New York.

Schmoldt, D.L., and H.M. Rauscher. 1995. *Building Knowledge-Based Systems for Natural Resource Management.* Chapman-Hall, New York.

Senge, P.M. 1990. *The Fifth Discipline: the Art and Practice of The Learning Organization.* Currency Doubleday, New York.

Index to Volumes I–III

abundance, I: 81; II: 47, 51, 54–56, 75, 140, 180
Acadia National Park, Maine, II: 213
accelerator mass spectrometry (AMS), II: 481
accounting, full-cost, I: 181
acid rain, I: 232; II: 173, 404, 698; III: 334, 605
acidification, II: 228, 233; III: 573, 577
Across Trophic Level System Simulation (ATLSS), II: 71–72
adaptation, I: 212; II: 25; III: 388, 513, 515–516
adaptive management, I: 7, 33, 37, 95, 98, 110, 131, 207–214,
    216, 243; II: 5, 10, 33, 125–126, 248, 262, 273, 275–276,
    319, 407, 522, 541, 591, 609, 625; III: 9, 14, 22, 28, 33, 324,
    419, 448–449, 505–534, 536, 552, 565–566, 692
Adaptive Management Area (AMA), III: 255
Administration Procedures Act, III: 34
adoption program, III: 422, 478
Advanced Research Projects Agency (ARPA), II: 187
Advanced Very High Resolution Radiometer (AVHRR), III:
    642–643, 655
African American, III: 191
agency
— federal, I: 80
— governmental, II: 15
— public, I: 123, 124
agency behavior, I: 123–127, 148; II: 323; III: 29, 85
agreement, I: 147; III: 117, 161
— implementation, I: 132
agreement seeking, III: 97, 100, 118–127
Agricultural Research Service (ARS), II: 424
agriculture, I: 63, 77; II: 26, 30, 72, 74, 171, 321, 434, 439, 501,
    532, 588, 653; III: 332
— sustainable, II: 260; III: 173
Alaskan National Interest Lands Conservation Act
    (ANILCA), II: 535; III: 190, 222, 237
Alaskan Native Claims Settlement Act, II: 535
alder, red, III: 523
algae, II: 59, 585; III: 75
all terrain vehicle (ATV), II: 631; III: 270
Allee effect, II: 98, 99–100, 104
— alliance, II: 364, 422, 425
alligator, II: 258
amenity value, III: 308
American Farm and Ranch Protection Act, III: 466
American Forest and Paper Association, II: 7

American Indian Religious Freedom Act (AIRFA), I: 146;
    III: 190, 204, 211
American Indian see Native American
American Industrial Heritage Project (AIHP), III: 80
American National Standards Institute (ANSI), III: 615, 640
American Society for Testing and Materials (ASTM), III:
    640
analysis of variance (ANOVA), II: 68; III: 543, 701
Angeles Restoration Crew (ARC), III: 220
animal unit month (AUM), I: 82, 182; II: 568; III: 230, 379
Annual Vegetation Inventory and Monitoring System
    (AVIMS), III: 657
ant, II: 22
— red fire (*Solenopsis invicta*), II: 589
antelope, pronghorn (*Antilocapra americana*), II: 124, 447,
    487, 568, 622; III: 334
anthracnose, dogwood, II: 637
anthropology, I: 151; III: 232
Antiquities Act, III: 11
— Apache-Sitgreaves National Forest, Arizona, II: 453; III:
    520
aphid, hemlock, II: 304
Applegate Partnership, Oregon, I: 140, 162, 163; III: 140,
    147, 157, 162–163, 252–258, 514
approach
— bottom-up, II: 364, 588; III: 80
— chicken-little, III: 666, 668
— coarse-filter, II: 41, 324, 333, 612; III: 438, 556, 647, 665
— do nothing, III: 675
— fine-filter, II: 324, 333, 612; III: 438
— functional, II: 45–86
— individual differences, III: 674
— integrated, I: 57; III: 247
— least-cost, I: 177, 182
— reactive, I: 209; III: 510–511
— top down, III: 609
Aquatic Habitat Classification System (AHCS), II: 408–410
aquatic reserve system, II: 573
aquatic unit, II: 368
Archaeological Resources Protection Act (ARPA), I: 74; II:
    495; III: 211
archaeology, I: 71–72; II: 471, 473, 479, 482, 484–487, 493,
    495, 497–499, 508

Archaic period, I: 74; II: 500–501
Arches National Park, II: 596
Area of Critical Environmental Concern (ACEC), II: 654–655
arrowhead (*Sagittaria* spp.), II: 451
Arroyo Cuervo region, New Mexico, I: 73
artefacts, cultural, I: 71, 72, 75; II: 493
artificial intelligence, III: 708
Ashley National Forest, Utah, II: 407
Asian American, III: 191
Asian Pacific American Employee Association (APAEA), III: 216
aspen (*Populus tremuloides*), I: 97; II: 442, 577; III: 511
aspen, quaking, II: 630
assessment, joint, I: 131
assessment contract, III: 546
assessment design, III: 538–546
assessment management, III: 538
assessment method, I: 215–221
assessment product, III: 546
assessment scale, I: 237; II: 415
assessment type, I: 237
association, II: 364, 370, 422
Augusta Creek landscape design (ACLD), II: 573
authority system, I: 174; III: 263, 266
autosuccession, II: 236

bacteria, II: 49, 51, 59, 66, 231
balance concept, II: 548, 586; III: 77, 401
bank stabilization, II: 641–642, 645–646
bark beetle, II: 289, 454
baseline condition, I: 71, 92; II: 264
baseline data, I: 235; II: 184, 207, 402
basin morphometry, II: 168
basketweaver, II: 450–451
bass, large-mouth, II: 585
bat, Indiana, III: 479
Bayesian belief network (BBN), II: 142; III: 419, 708
bear
— black, I: 130; II: 198; III: 112
— brown, I: 130; II: 198; III: 112, 125
— grizzly (*Ursus arctos horribilus*), I: 48, 79, 109; II: 12, 119, 124, 125, 148–150, 198, 201, 293, 446, 640; III: 15, 438
— short-faced (*Arctodus simus*), II: 446
beaver (*Castor canadensis*), I: 31, 37, 157; II: 71, 80, 180, 258, 446, 643, 646, 648; III: 238, 511
beaver, giant (*Castoroides ohioensis*), II: 446
bee, II: 56
beech (*Fagus grandifolia*), II: 316, 442
Best Available Retrofit Technology (BART), III: 124
best management practice (BMP), I: 192; II: 72, 636, 647; III: 425, 446, 451, 469, 485, 510
bioclimate, II: 202
biodiversity, I: 13–18, 78, 79, 178, 186, 199; II: 4, 7, 8, 13, 39–41, 47, 56, 72–73, 157, 169–170, 198, 212, 274–275, 290, 334, 519, 527, 539, 548, 638, 699; III: 363, 397, 475, 515, 545, *see also* diversity and species diversity
— pattern approach, II: 374
— trust fund, III: 474

biodiversity conservation, I: 58, 208; II: 73, 355, 377, 398, 532; III: 13, 18, 165–167
biodiversity loss, III: 184, 333–334
biodiversity network, II: 550
biogeochemical process rate, II: 160
biogeography, I: 20; II: 353, 355
biogoecoenosis, II: 370
biological control, I: 99; II: 304, 638
biological diversity *see* biodiversity
Biological Resource Division (BRD), II: 186
biomass, I: 15; II: 47, 50, 55, 65, 180, 294, 530–531; III: 537
biomass transfer, II: 235
biome, II: 238
bioremediation, II: 602
biosociology, III: 186, 241, 285
biosphere, II: 333
biotechnology, III: 335
biotemperature, II: 238–239
biotic integrity index, II: 403
birch (*Betula* spp.), II: 442
— yellow (*Betula alleghaniensis*), II: 316
birding, III: 373
birth pulse, II: 94
birth rate, II: 119, 679, 681
bison, American (*Bison bison*), I: 21; II: 61, 80, 111, 116, 257, 444, 446; III: 464
bison, big-horned (*Bison latifrons*), II: 446
bitterroot (*Lewisia rediviva*), II: 487
Bitterroot National Forest, Montana, II: 487, 525, 526
black box, III: 307, 526
black grass (*Juncus gerardii*), II: 610
blackbird, II: 75
blight, chestnut, I: 16; II: 55, 304; III: 334
blue dick (*Dichelostemma capitatum*), II: 453
bobwhite quail, Northern (*Colinus virginianus*), II: 116, 124, 175, 316
boundary, I: 219; II: 247, 336
— administrative, II: 344; III: 541, 553
— ecological, II: 4, 41, 376
— ecoregional, I: 219
— political, II: 41, 200, 559; III: 300
boundary movement, II: 182
boundary permeability, II: 182–183
bracken fern (*Pteridium aquilinum*), II: 453
Braun–Blanquet method, II: 362, 364
Breeding Bird Survey (BBS), II: 210, 415
breeding birds, II: 530
Bridger-Teton National Forest, III: 654–655
Brundtland Report, II: 39
Bryce Canyon National Park, III: 395
budworm, spruce, II: 289, 293, 656
buffer zone, I: 32; II: 213, 272, 447, 611, 645, 647, 650; III: 470, 477, 524
bullfrog (*Rana catesbeiana*), II: 80
Bureau of Census, III: 54, 64, 338
Bureau of Indian Affairs (BIA), III: 195, 215
Bureau of Land Management (BLM), I: 109; II: 5, 181, 206, 423, 424, 518, 537, 559, 650, 654; III: 3, 19, 196, 201, 215, 269, 348, 588, 650

Bureau of Reclamation (BOR), II: 6, 319
bureaucracy, I: 123, 126, 135, 148, 161; III: 136, 184, 248, 257
bureaucracy culture, I: 89; III: 201, 215
burning
— frequency, II: 445
— prescribed, I: 64, 147, 191, 194; II: 79, 317, 454, 462–464, 486, 567, 605, 632; III: 197, 218, 450
business process re-engineering (BPR), III: 633
butterfly, Karner blue, III: 525

cactus, Saguaro (*Carnegiea gigantea*), II: 53
California Biodiversity Council (CBC), III: 165–167
California Desert Act, III: 196
California Environmental Project (CEP), III: 220
California Environmental Resources Evaluation System (CERES), II: 186; III: 167
California Wildlife Habitat Relationship (CWHR), II: 417
camas (*Camassia quamash*), II: 450, 487
camel (*Camelops hesternus*), II: 446
Canadian Forest Inventory Committee (CFIC), II: 418
Canadian National Forest Inventory (CanFI), II: 417–418
canonical correlation analysis, III: 700
Cape Hatteras National Seashore, North Carolina, II: 208
Captive Breeding Specialists Group (CBSG), II: 125
carbon cycle, II: 50, 228–230
carbon dioxide, II: 50, 74, 76, 227, 229–230, 263, 698; III: 333
carbon sequestration, I: 199, 201; III: 397, 399, 468, 474–475
carbon sink, I: 187; II: 698
carbon to nitrogen ratio, II: 63, 229, 231
carbon to phosphorus ratio, II: 229
caribou (*Rangifer tarandus*), II: 447
carnivore, II: 50, 58, 71, 79, 80, 228, 447
carp, I: 152
carrying capacity, I: 154; II: 23, 43, 77, 90, 93, 97–98, 104–105, 113–114, 122–123, 446, 524, 679; III: 230, 401
Carson National Forest, New Mexico, III: 199, 213
cartography, III: 613
case studies, I: 70; II: 5, 42, 146–152, 257–261, 264–273, 275, 318–319, 402–403, 405–410, 412–418, 424–425, 484–490, 561–579, 631, 633, 638, 644, 646, 652, 654–655, 658, 660; III: 52, 111–115, 123–127, 160–178, 193–196, 199–201, 215–223, 252–274, 365–380, 393–398, 440–444, 450–453, 457–459, 580–581, 597, 619–626, 646–658, 692–697, 716–717
caste, II: 31
cat, sabre-tooth (*Homotherium* spp.), II: 446
cation exchange capacity, II: 51
cattail, narrow-leaved (*Typha angustifolia*), II: 453, 611
cedar, Lebanon, II: 26
census, I: 238; III: 59, 645
census data, I: 118; II: 108, 122; III: 64, 66
challenge cost-sharing, III: 117
change
— environmental, I: 15; II: 53, 161, 383, 474
— legal, III: 33
— scientific, III: 28
— social, III: 63
— technological, II: 686
change indicator, III: 565

channelization, I: 86; II: 652
chaparral, I: 41, 42; II: 78, 288, 462, 586
charcoal, II: 325, 477, 497
Chattahoochee National Forest, Georgia, II: 324
cheatgrass, Eurasian (*Bromus tectorum*), II: 52, 76, 299, 304, 632, 637
Chequamegon National Forest, Wisconsin, II: 402–403; III: 590
Chesapeake Bay, I: 187; II: 212–213
chestnut, American (*Castanea dentata*), I: 16; II: 51, 80, 530; III: 282, 334
Chugach National Forest, Alaska, II: 202
Chicago School, II: 24
chicken, prairie, II: 113
Chippewa National Forest, II: 423
chuckwalla, III: 194
Cibola National Forest, New Mexico, III: 213
citizen, I: 123; II: 15; III: 516
citizen advisory committee, III: 173
city-dwelling, II: 682
Civil Rights Act, III: 210
clan, II: 31
class, II: 31
— successional, II: 67
— taxonomic, II: 396
classification, I: 34, 55; II: 123, 368–375, 420, 623 *see also* ecosystem classification
— abiotic, II: 356
— application, II: 375–383, 395–432
— aquatic, II: 366, 368–369, 383
— biogeographic, II: 356
— biotic, II: 356
— channel-type, II: 271
— climatic, II: 360
— coastal, II: 368, 383
— community, I: 155
— ecological, I: 27, 53–60, 139; II: 43, 209, 361, 363–375 *see also* ecological classification system
— ecoregional, II: 370–375
— emic, I: 155, 156
— etic, I: 156, 157; III: 239
— floristic, II: 360, 364
— freshwater, II: 366–367
— global, II: 357
— hierarchical, II: 356, 365, 397, 420
— horizontal, I: 155; III: 234–237
— key attributes, II: 377–379
— land cover, II: 375
— landscape, I: 139
— limits, I: 152
— multi-factor, II: 357, 370, 405
— physiognomic, II: 364
— single-factor, II: 357, 369
— social, I: 151–158; III: 185, 227–244
— terrestrial, II: 363–366, 383, 405–407
— value-laden, III: 228
— vertical, I: 154; III: 232–234
— zoogeographic, II: 365
Classification and Multiple Use Act, III: 3

Clean Air Act, I: 109; II: 6, 464, 525, 530, 540; III: 11, 124–125
Clean Water Act, I: 108, 109, 110; II: 6, 525, 530, 540; III: 11, 24, 450–451
clearcutting, I: 31, 41; II: 182, 204, 525, 526, 590, 694, 697; III: 74, 280–281 *see also* logging
— checkerboard, I: 41; II: 291
climate, I: 15, 30, 353; II: 71, 230, 446
climate change, I: 31, 43, 44, 69, 208; II: 31, 75–76, 299, 303, 315, 698; III: 327, 333, 397
Climate Change Convention, III: 331
climate warming, I: 232; II: 7, 182, 263, 611, 692; III: 474, 605
climax, II: 298, 358, 359, 364, 369, 529, 549; III: 664–665
climbing, recreational, III: 198
clover (*Trifolium* spp.), II: 453
cluster analysis, II: 202, 362; III: 623, 700
co-evolution, I: 71; II: 494
coastal system, I: 57; II: 73, 355, 366, 652
Coastal Zone Management Act, II: 6
Cocoino National Forest, Arizona, II: 576
coexistence, II: 40, 101, 122
coffee bean, II: 26
cogongrass (*Imperata cylindrica*), II: 589
cold war conservation, III: 92–94
collaboration, I: 79, 129–134; II: 10; III: 3, 5, 97–106, 128, 131, 134, 141, 148, 306–307, 528, 588
— interagency, I: 4, 133, 219; III: 268
colluvium, II: 172
colonization, I: 20, 29, 65; II: 110
Colville National Forest, II: 578
Commercial Off The Shelf (COTS), III: 641
Committee for European Normalisation (CEN), III: 640
commodity, I: 81, 177, 186; II: 5, 27, 243, 517, 686; III: 196, 346
commodity market, I: 178; III: 401
commodity output, II: 533, 539
commodity production, I: 224; II: 526, 561; III: 565
commodity value, III: 308
common pool resource (CPR), III: 131, 138
Common Survey Data Structure (CSDS), II: 423
commons, III: 138, 263
— tragedy of, III: 135, 365
Commonwealth Agricultural Bureau (CAB), III: 49
communication, I: 159, 161, 209, 243; II: 355; III: 46, 134, 250, 261, 665
— cross cultural, I: 233; III: 189, 196, 198–201, 210, 215, 224
— direct, III: 610, 616
communication system, I: 245; II: 9; III: 260, 681
community, I: 159; II: 15, 67, 180, 359, 363; III: 175, 233, 297
— of expertise, III: 251
— Hispanic, I: 150
— human, I: 195
— of interest, I: 162; III: 107, 251, 257, 268, 273, 309, 531, 630
— international, II: 7
— organic, III: 251
— of place, I: 162; III: 251, 253, 257, 273, 531
— potential natural (PNC), II: 425
— rural, II: 25, 551
community assessment, III: 254

community decision-making, III: 189, 205
community ecology, III: 282
community involvement, III: 203, 213, 255
community model, II: 70–71
community plan, III: 107
Community Re-investment Act, I: 200; III: 479
community response, II: 235
community stability, II: 537; III: 163, 545
community sustainability, II: 575
competition, I: 15; II: 56, 63–65, 69, 70, 98–99, 101, 300, 588; III: 88, 216, 299
— global, I: 182; III: 366
competition model, II: 101–102
competition theory, III: 287
competitive exclusion principle, II: 62
complexity, I: 36, 238; III: 103–104
— ecological, I: 36; II: 222
complexity management, III: 31–32
complexity theory, II: 315
compliance monitoring, III: 589
component
— abiotic, II: 360–361
— biotic, II: 359–360
composition indicator, II: 265–266
Comprehensive Environmental Response, Compensation and Liability Act (CERCLA), III: 24–25
computer-aided systems engineering (CASE), III: 633
concern identification, III: 110
condition
— biophysical, II: 8, 13
— ecological, I: 50
— environmental, I: 15; II: 72, 360
— existing, I: 237
— natural, I: 57
— paleoclimatic, II: 480
— pre-settlement, II: 433, 473
condor, California, II: 113
conflict, I: 159–160, 164; III: 247–248, 268, 272, 286
— bounded, III: 554, 559
— management, III: 99, 140, 157, 178, 300, 310
— sociocultural, III: 97–98, 185, 256
conflict resolution, III: 156
Conjoint Analysis (CA), III: 392
connectivity, II: 162, 164, 176, 182, 339
— hydrologic, II: 652
consciousness development, III: 47, 134
consensus, I: 129; III: 97, 106, 118, 126, 159, 670
consensus building, I: 132; III: 99, 122, 253, 676, 690
consequence assessment, III: 436, 438
conservation, I: 17, 115, 126
conservation agreement, I: 198; III: 467
conservation banking, III: 478
conservation biology, I: 96; II: 8, 201, 418, 455, 528, 539; III: 10
conservation easement, I: 197–198, 200; II: 213; III: 466, 469
conservation group, III: 113
conservation movement, II: 523; III: 88
conservation network, II: 533–534, 539
Conservation Options and Decisions Analysis (CODA), II: 202

conservation planning, II: 48
conservation policy, II: 524
conservation reserve program, III: 481
conservation strategy, II: 628
conservation value, I: 100; II: 628; III: 397, 478
conservation-of-mass equation, II: 229
constraint, I: 37, 86, 99; II: 42
— natural, II: 690–691
consumption, I: 171; II: 11, 27, 517, 677
— per-capita, I: 82
consumption behavior, I: 198; III: 61, 329
contagion, II: 162, 286
contaminant, II: 657 *see also* pollutant
continent, III: 310
contingent valuation (CV), III: 323, 391, 394–398
continuity, functional, II: 221
continuum concept, II: 359, 363
control
— inefficiency, III: 398
— technocratic, I: 126
controlling factor approach, II: 68, 370–371
cooperation, I: 136–137; III: 5–6, 134
— barrier, II: 419; III: 144
— incentive, II: 419
— institutional, I: 138
— interagency, I: 79; III: 115, 156
— promotion, III: 139
— regional, I: 135–140; III: 155
— voluntary, III: 169
Cooperative Weed Management Area (CWMA), II: 206
coordinated resource management (CRM), II: 566
coordination, III: 134, 305
— interagency, I: 81
— interjurisdictional, III: 13
coppicing, II: 632; III: 194
cordgrass
— salt meadow (*Spartina patens*), II: 610
— saltwater (*Spartina alterniflora*), II: 610
corridor, II: 162, 164, 258
— migratory wildlife, III: 21
corridor index, III: 623–624
cost, I: 179, 185, 189, 194, 236; III: 385
— direct, III: 435, 445
— economic, I: 80
— legal, III: 435
— marginal, III: 355
— of transaction, III: 387
— total cost of risk (TCR), III: 449–450
cost–benefit analysis, I: 89, 99, 189; II: 11, 404; III: 18, 363, 373, 385, 389, 393, 704
cost effectiveness, I: 100; II: 561, 594, 629–630, 638; III: 449
cotton, II: 315
cottonwood (*Populus* spp.), II: 453, 642, 649, 654; III: 194, 581
cougar, II: 60
Council on Environmental Quality (CEQ), I: 174; III: 21, 336, 552
covenant, I: 198; III: 467
cowbird, brown-headed (*Molothrus ater*), II: 334

crane, whooping, III: 142
cress, hoary, II: 637
crisis management, III: 510, 667–668 *see also* chicken little approach
cropland area, II: 684–865
Cross-Cutting Act I: 108
crown fire, I: 40, 42; II: 296
culture, I: 169; II: 677, 698
— American, III: 156
— as dynamic system, I: 146
— declining, III: 192–193
— indigenous, III: 193
— institutional, III: 145
Customer Use and Survey Techniques for Operations, Management, Evaluation and Research (CUSTOMER), III: 73
cycle
— biogeochemical, I: 31, 34; II: 31
— continuous, I: 224
— environmental II: 31
— social, II: 24, 25, 30–31

damage
— environmental, III: 350
— riparian, III: 253
data, I: 22, 215, 217, 231; II: 27; III: 637
— archaeological, I: 17; II: 476–480, 497–499; III: 606
— demographic, II: 143; III: 700–701
— dendrochronological, I: 98; II: 506
— empirical, III: 45
— existing, I: 228; III: 593, 640, 646
— experimental, II: 68
— fine-scale, II: 337
— functional, I: 238
— geological, I: 98
— geospatial, I: 238–239; III: 553
— half-life, III: 606
— historical, I: 92; II: 273, 287, 316, 482, 486–487, 594, 604; III: 573, 578, 606
— long-term, II: 12
— paleobotanical, I: 67, 98; II: 477, 490
— paleoecological, II: 475
— paleoenvironmental, I: 76; II: 497
— physical, I: 17
— poor, I: 225
— primary, I: 118; III: 63, 67
— qualitative, I: 120; III: 63, 77
— quantitative, III: 63
— raw, III: 63
— secondary, I: 118, 120; III: 49, 63, 66, 76, 235 *see also* existing data
data accuracy, III: 636
data analysis, I: 139, 220; II: 68; III: 53, 543–545
data availability, III: 617–618, 623, 638
data collection, I: 139, 231–239; II: 625; III: 53, 76–78, 501, 545, 593–596, 603–628, 631, 643–646
data communication, III: 605, 607
data dissemination, II: 186–187
data format, III: 617–618

data management, I: 231–239; III: 587, 603–628
data ordination, II: 362
data ownership, III: 607
data preservation, I: 228, 232
data quality, II: 507; III: 370, 636
data requirement, I: 237; II: 118
data sets, Government, III: 64–65
data sharing, III: 631, 635–641, 653
data source, III: 617–618
data standards, I: 238–239; III: 637
data technology, I: 238
data type, II: 375; III: 617–618
data use, I: 232; II: 66–67; III: 608
database, III: 595, 610
— compatibility, III: 557
— computer-based, III: 49
— digital, II: 424
— relational, III: 609, 614
— standardized, II: 423
— subject-matter, I: 236; III: 631
database management, III: 502, 614–615
database program, II: 399
dating method, II: 480
death rate, II: 103, 119, 679
Death Valley, III: 193–196
decision analysis, III: 633
decision appraisal, III: 457
decision maker, I: 216, 242, 243; III: 670
decision making, I: 5, 136, 193, 226, 242, 243, 245; II: 8, 348,
    592; III: 297, 510, 662, 689, 709
— cooperative, II: 461; III: 133
— iterative, II: 248; III: 137
decision making method, III: 667–669
decision making process, II: 5, 13, 30, 321; III: 61, 289, 421,
    688, 692–697
decision space, III: 102, 110, 505, 507
decision support system (DSS), I: 241–246; II: 335; III: 279,
    289–292, 661–686, 705, 711–718
— ArcForest, III: 712
— ArGIS, I: 244; III: 712
— CRBSUM, III: 712
— EMDS, III: 558, 675, 712
— Forest Vegetation Simulator (FVS), III: 712
— GypsES, III: 713
— INFORMS, I: 245; III: 713
— KLEMS, III: 713
— LANDIS, III: 713
— MAGIS, III: 714
— Northeast Decision Model (NED), I: 245; III: 279,
    291–292, 675, 714
— RELMdss, III: 714
— SARA, III: 714
— SIMPPLLE, III: 715
— SNAP, III: 715
— SPECTRUM, I: 244; III: 715
— Terra Vision, III: 715
— Woodstock, III: 716
decision support tool, III: 688–708
decision tree, I: 119, 120, 243; III: 48

decline, ecological, I: 96
decomposer, II: 49, 50, 228; III: 282
decomposition, II: 49, 50, 230, 266
deed restriction, I: 198; III: 467
deer (*Odocoileus* spp.), II: 71–72, 75, 76, 80, 447
— mule, III: 605
— white-tailed (*Odocoileus virginianus*), I: 48; II: 81, 175,
    500, 531, 578, 622; III: 334, 511
deforestation, II: 7, 177, 480, 506, 684, 694; III: 332, 402
— tropical, II: 865
deforestation rate, II: 686
degradation, I: 95–96, 181; II: 205, 604, 656, 679
Delphi technique, II: 410; III: 52–53, 145, 435, 439, 681, 704
demography, I: 151, 169–175; II: 23, 135, 677, 692; III:
    328–331, 691, 702
Demonstration of Ecosystem Management Options
    (DEMO), II: 634
Denaina plant classification, III: 238
Denali Mine Reclamation, II: 325–326
Denali National Park, II: 324
dendrochronology, II: 478, 605
dendroclimatology, II: 478
denitrification, II: 232
Department of Environmental Regulation (DER), III: 145
Department of Natural Resources (DNR), II: 574
Department of the Interior (DOI), III: 201
dependency theory, II: 679
Deschutes National Forest, II: 636
Desert Lands Act, III: 10
desertification, II: 181, 632
— indicator, II: 637
desired future condition (DFC), I: 125; II: 4, 313, 320, 395,
    404–410, 412, 538, 573, 592, 624, 633, 659, 663; III: 12, 30,
    62, 94, 110, 158, 163, 213, 380, 437, 508, 552–553, 555, 697
detritivores, II: 228
Devils Tower National Monument, I: 148
diameter at breast height (DBH), II: 414
digital elevation model (DEM), II: 71; III: 654
Digital Orthophoto Quadrangles (DOQ), III: 654
diking diversion, II: 653
dipper (*Cinclus cinclus*), II: 124
discount rate, III: 395
discounting, III: 407
disease, I: 15; III: 17, 334
— European-introduced, II: 434, 440–441, 503
— prevention, I: 194
disintegration, physical, II: 165
dispersal, II: 58, 69, 91, 96, 99, 107–109, 201
dispersal corridor, II: 631
displacement, competitive, II: 65–66
disposition era, III: 10
dispute resolution, environmental, III: 99
distribution, spatial, I: 170, 181
disturbance, I: 30, 33, 39–41, 98; II: 41–43, 60, 66, 167,
    171–173, 195, 203–207, 219–220, 235, 281–313, 316–317
— amplification, II: 204
— anthropogenic, I: 41; II: 77, 303, 314, 320, 471, 587
— aquatic, II: 172–173
— environmental, I: 15

— feedback, II: 281, 286, 288
— geologic, II: 171–172
— historical, I: 43
— local, I: 179
— magnitude, II: 287
— mortality-causing, I: 16
— multiple, I: 41
— natural, I: 41, 44; II: 64, 77, 169, 205, 283–284, 314, 320, 548, 586
— physical, II: 222
disturbance dynamics, I: 39–46; II: 40
disturbance ecology, II: 8, 283, 302, 313–330
disturbance frequency, II: 64–66, 77, 79, 292, 297–298
disturbance history, I: 41
disturbance initiation, II: 293
disturbance interaction, II: 281, 290
disturbance management, II: 77–79
disturbance patch, I: 86
disturbance process, III: 555
diversity, II: 162
— alpha, II: 56
— beta, II: 56, 62, 374
— between-habitat, II: 56–57, 62
— ecological, II: 169; III: 692
— genetic, I: 13–18; II: 27, 39, 46, 48, 115, 167, 529; III: 522 see also biodiversity and species diversity
— global, II: 532
— landscape, I: 27–32
— local, II: 56
— regional, II: 56–57
— sociocultural, I: 145–150; III: 184, 189–208, 210
— socioeconomic, III: 257
diversity index, III: 623–624
diversity pattern, II: 45
diversity–productivity curve, II: 63
documentation, I: 212, 243; II: 481–483; III: 599, 633
Douglas-fir (Pseudotsuga menziesii), II: 173, 221, 267, 289, 302, 454, 574, 607; III: 374, 509, 523
drainage, III: 309
— acid, I: 101; II: 275, 660, 657
drainage network, II: 172
drainage structure, II: 652
drought, II: 31, 302, 655; III: 199
dune stabilization, II: 206
Dutch elm disease, II: 55
dynamics, temporal, I: 39–46; II: 45, 281–312

eagle, bald, II: 654
early warning, I: 92, 227; III: 564
earthquake, II: 225
earthworm, I: 35; II: 49, 231
ecoclimatic region, Canada, II: 371
ecolabeling, I: 201; III: 472 see also green product labeling
ecological assessment, I: 244; III: 542, 555
Ecological Classification and Mapping (ECOMAP), II: 371–374, 408, 425
ecological classification system (ECS), II: 402; III: 170 see also classification
ecological economic stewardship, III: 404

ecological economics, I: 185–190; II: 8, 337; III: 55, 383–412
Ecological Land Type (ELT), II: 402–403, 412 see also land type
Ecological Land Type phase (ELTP), II: 402–403, 407, 412
ecological reporting unit (ERU), III: 368–371
Ecological Society of America (ESA), II: 357, 600; III: 229, 438
ecological stewardship, I: 1, 4, 197–204; II: 4
ecological unit, I: 55; II: 395, 398, 421; III: 654–655
ecological theory, I: 54; II: 358
ecological thinking, II: 548–550
ecologist, II: 14, 691
ecology, I: 179, 185; II: 257, 274; III: 310
economic activity, I: 68
economic analysis, I: 177, 189
economic arrangement, II: 692
economic benefit, III: 361, 371
economic control, II: 537
economic development, II: 680
economic growth, II: 39; III: 329–330, 393
economic health, II: 538; III: 405
economic impact analysis, I: 192–204; III: 373, 376
economic theory, III: 61
economic tools, III: 324
economic value, III: 373–374
economics, I: 169, 170, 179, 186; II: 503, 677, 679; III: 346
— general equilibrium, I: 179; III: 346
— neo-classical, I: 186; II: 679
— partial equilibrium, III: 346
economy, I: 185; II: 29
— industrial, III: 250
ecoregion, I: 53, 55, 120, 215, 353; II: 33, 356, 370, 374; III: 161
— United States, II: 371
ecoregional assessment, I: 216, 219; III: 62, 552, 559
ecosystem, II: 4, 7, 13, 23, 219, 221–223, 303, 315; III: 282, 664
— aquatic, I: 55; II: 355, 421; III: 24
— compartment, II: 588
— component, II: 358–361, 622; III: 539, 567
— constraint, I: 37
— natural assets, I: 186
— network, II: 504
— property, I: 89, 90; II: 50–51, 358, 361, 584, 589
— terrestrial, I: 56, 96
ecosystem analysis, II: 42, 196, 343–352; III: 238
ecosystem approach, II: 4, 157–194; III: 144, 266
ecosystem assessment, I: 215; II: 623
ecosystem attribute, II: 396
ecosystem behavior, II: 13, 125, 223, 376; III: 286, 573, 580
ecosystem boundary, II: 344; III: 539
ecosystem change, I: 227; II: 12–13, 273–276, 454, 586
ecosystem complexity, I: 34, 209; II: 8, 22, 336, 396, 591; III: 306, 384, 565
ecosystem composition, II: 39–41, 401
ecosystem condition, I: 85–94, 179, 220; II: 583; III: 555, 577
ecosystem conservation, II: 4
ecosystem disturbance, I: 29, 35, 85; III: 537
ecosystem diversity, I: 209, 17, 27–32; II: 336, 340, 408–409, 414

ecosystem dynamics, II: 8, 22, 281, 549, 587, 623; III: 537, 580

ecosystem flux, II: 246

ecosystem function, I: 33–38, 85, 97, 188, 217; II: 8, 39–41, 49–51, 241, 244, 219–254, 255–280, 359, 401, 436, 564, 587; III: 536

ecosystem goods, I: 212

ecosystem health, I: 78, 79, 81, 185, 186; II: 53, 198, 206, 400, 519, 557; III: 197, 384, 406, 576–577, 594

ecosystem heterogeneity, II: 79

ecosystem history, II: 496

ecosystem integrity, I: 40, 79, 124, 228; II: 260, 262; III: 16, 544–545

ecosystem interface, II: 240

ecosystem management (EM), I: 1, 3–7, 78, 79–81, 83, 87, 179, 188, 242; II: 3–20, 219, 244, 262, 326, 520; III: 31, 103, 162, 172, 298, 361, 558 see also natural resource management

Ecosystem Management Decision Support (EMDS), I: 244; III: 558, 675, 712

ecosystem management, approach, I: 2, 48, 89, 188; II: 246–248, 262, 559–561

ecosystem management, principles, II: 34, 44, 245; III: 12–14

ecosystem medicine, II: 601

ecosystem model, III: 571

ecosystem monitoring, I: 227; III: 572

ecosystem output, I: 178, 179, 181; II: 627; III: 360, 366, 370–372

ecosystem planning, II: 411–416; III: 235

ecosystem process, I: 2, 33–38, 39, 49, 90, 217, 220, 225; II: 40, 219–254, 255–280, 657; III: 89, 536

ecosystem product, I: 181

ecosystem productivity, II: 359, 572

ecosystem recovery, II: 283, 602; III: 592

ecosystem resilience, I: 85, 86, 87, 90, 185; II: 52, 243, 304, 436; III: 287, 384, 544–545 see also resilience

ecosystem response, I: 41, 212; II: 223, 287, 300–302, 410, 585; III: 565

ecosystem restoration, II: 351, 547, 576, 612, 633

ecosystem scale, II: 41

ecosystem service, I: 13, 179, 181, 185, 190, 212; II: 11, 25, 26, 73, 517, 536; III: 346, 393, 407

ecosystem stability, II: 52, 629

ecosystem stress, I: 227; II: 629; III: 575, 577

ecosystem structure, I: 13, 40, 85, 90, 97, 217; II: 39–41, 49, 400–401, 436; III: 536

ecosystem sustainability, I: 85–94, 89, 124, 212; II: 74, 563, 583–598, 592; III: 16, 500, 507, 692 see also sustainability

ecosystem threshold, II: 315, 321

ecosystem trajectory, I: 224

ecosystem transformation, II: 528

ecosystem value, I: 78; III: 564

ecosystem variability, I: 87; III: 592

ecotone, II: 168, 358, 376

ecotourism, II: 699; III: 333, 373

Ecotrust, III: 473

ecotype, I: 53, 55

ecozone, II: 210, 371, 418

ectomycorrhizae, II: 231

edge effect, II: 163, 170

edge habitat, II: 79

edge number, II: 162

education, II: 29, 638; III: 47, 168, 216, 454, 622

— environmental, III: 174, 211

education program, III: 555

effect, cumulative, I: 51, 87, 90; II: 204–206, 346, 355, 411, 474, 585, 626; III: 14, 21–22, 33, 362, 404

effectiveness measure, III: 527–529

effectiveness monitoring, III: 31, 589–590

efficiency, II: 49, 50; III: 385

— dynamic, III: 386

— economic, II: 699; III: 665, 679

— production-line, III: 86

— static, III: 385–386

eigenvalue, II: 95–96

eigenvector, II: 95–96

El Nino, II: 12, 31

elasticity, II: 96, 144, 223; III: 437

elderberry (Sambucus mexicana), II: 452

elephant, I: 35

elevation, II: 63

elevation effect, II: 170

elk (Cervus elaphus), II: 76, 111, 337, 446–447, 577, 622; III: 334, 374, 510–511, 590–591

elm, American, II: 80, 530

embeddedness, social, III: 234

emission, III: 335

emission allowance, III: 400

employment, III: 351

Endangered Species Act (ESA), I: 6, 21, 23, 107, 108, 109, 110, 190, 198, 201 II: 5, 6, 7, 9, 29, 30, 40, 46, 137–138, 145, 196, 200, 521, 525–526, 527, 540, 561, 607, 640; III: 4, 9, 11, 13–15, 22–24, 33–34, 142, 150, 201, 270, 352–353, 389, 440, 450–451, 464, 476, 513, 520, 677

endemism, II: 54, 360

endomycorrhizae, II: 231

endpoint assessment, III: 418, 433, 437, 553

energy, II: 25–26

— nuclear, II: 26, 28

— solar, II: 26, 49, 74, 223

energy analysis, III: 407

energy budget, II: 229

energy consumption, II: 70, 696

energy cost, III: 407

energy efficiency, I: 34

energy flow, I: 34, 188; II: 25–26, 65–66; III: 282

energy source, renewable, II: 693

energy transfer, II: 65, 256

engineering, ecological, II: 245

entropy, I: 126; II: 26, 223, 234; III: 91, 401

— organizational, I: 126; III: 91

environment, I: 170; II: 24

— aquatic, II: 167–168

— biophysical, I: 237

— regulatory, II: 137–138

environmental assessment (EA), II: 6, 210; III: 19, 30, 32, 115, 123, 676

environmental impact, I: 92

environmental impact statement (EIS), I: 111; III: 20, 30, 70, 113, 115, 271, 538, 676
environmental movement, II: 525–526, 543, 548, 558; III: 210
Environmental Protection Agency (EPA), II: 6, 420, 424, 658; III: 24, 142, 436, 484
Environmental Systems Research Institute (ESRI), III: 648
environmentalism, I: 116; III: 51, 54, 283
environmentalist, I: 202; III: 144, 162
epidemics, II: 116, 680
epiphyte, II: 241
equality, II: 13
equilibrium, I: 86; II: 223, 234, 294–295, 358, 548
— dynamic, II: 281, 294–300
— qualitative, II: 295
— quantitative, I: 43, II: 295, 299
— stable-trajectory, I: 43; II: 300
— stationary-dynamic, II: 296
— statistical, II: 296
— steady-state, I: 43; II: 223, 294–295, 298, 586; III: 401
erosion, I: 29, 35, 42; II: 59, 60, 220, 226
estate tax, I: 198; III: 471
estuarine classification, II: 367–368
estuarine-coastal system, II: 652
ethnoarchaeology, II: 498–500
ethnobotany, II: 439, 457, 459
ethnography, II: 439, 483–484; III: 51, 204, 288
ethnology, II: 483–484
ethnozoology, II: 457
European American, I: 73; III: 191
European settlement, II: 175, 319, 434, 604
eutrophication, II: 64, 75, 76, 233; III: 577
evaluation, I: 223–230, 245; II: 416–418; III: 105, 292, 523–525, 577, 598, 696
evapotranspiration, II: 51, 52, 71, 238
event tree analysis, III: 435
Everglades National Park, I: 95; II: 69, 76, 205, 284, 591; III: 35, 145, 592
— restoration, II: 255, 258–259; III: 176–178, 199
Everglades Nutrient Removal (ENR), II: 76
evolution, I: 63–70; II: 48, 201
evolutionary ecology, I: 96; II: 602
evolutionary environment, I: 98; II: 603
excavation, archaeological, II: 476
existence value, II: 699; III: 371, 379, 398
exoskeleton, II: 479
experience
— collective, II: 28
— on-the-ground, III: 610
— professional II: 8–9
experience transfer, II: 355
experimentation, I: 202, 207, 211; II: 15, 287, 414, 461; III: 523
expert judgement, III: 438
expert knowledge, III: 573
expert opinion assessment, II: 145–146
expert system, II: 203; III: 290, 688, 705, 708
expert, technical, III: 149
exposure, III: 416, 433
exposure assessment, III: 436
exposure intensity, I: 75

exposure pathway, III: 417–418, 433, 436
exposure threshold, I: 92; II: 594
externality, I: 173, 178; II: 5, 11, 699; III: 136, 300, 347, 352, 361–362, 387
extinction, I: 14, 19, 20, 21, 23, 44, 96, 190, 198; II: 27, 40, 46, 58, 60–62, 79, 81, 110, 245, 446, 529; III: 11, 23, 321
extinction model, II: 109–110
extinction rate, II: 48, 160, 335; III: 334
extinction risk, II: 167
extirpation, I: 23, 96; II: 6, 145, 258
Extremely Low Frequency (ELF), III: 591
Exxon Valdez, I: 157, 229; III: 239–240

facilitator, I: 126; III: 52, 109, 119, 124, 141, 157, 212, 439, 441, 693
factor
— biological, II: 179–184
— deterministic, I: 20
— direct, II: 360
— edaphic, II: 168
— environmental, II: 237, 401
— human, II: 384
— indirect, II: 360
— secondary, I: 20
— situational, I: 137
— stochastic, II: 114
factor analysis, III: 700–701
falcon, peregrine, II: 639
fan palm, desert (*Washingtonia filifera*), II: 452
farming system, II: 687–688
fauna, II: 26–27, 479
feasibility assessment, III: 115–116, 119, 709
fecundity, II: 96, 100
fecundity rate, II: 69, 94
Federal Advisory Committee Act (FACA), I: 107–109, 112, 140; II: 3, 5, 14, 200, 541; III: 4, 9, 13, 15, 19, 25–26, 36, 111, 118, 147, 159, 178, 255, 272, 671
Federal Cave Resources Protection Act, III: 271
Federal Energy Regulatory Commission (FERC), III: 123–124
Federal Environmental Pesticide Control Act, II: 6
Federal Geographic Data Committee (FGDC), I: 239; II: 185, 423; III: 639–640, 644
Federal Information Processing Standard (FIPS), III: 615
Federal Insecticide, Fungicide and Rodenticide Act (FIFRA), II: 525; III: 24
Federal Land Policy and Management Act (FLPMA), I: 108; II: 5, 6, 525, 535; III: 4, 9, 11, 19, 210, 440
fee setting, III: 174, 380, 400, 483–484
fee, user, III: 354, 471
fencing, II: 564, 567, 577, 629; III: 378
ferret, black-footed (*Mustela nigripes*), I: 224; II: 117; III: 136
fertilization, II: 49, 64, 74–75, 639
fertilizer, II: 26
field research, III: 50–54
Fifth Amendment takings clause, I: 112
financial capital, I: 186; II: 27
fir
— grand (*Abies grandis*), II: 607

— white (*Abies concolor*), II: 173, 454, 579; III: 194

fire, I: 39, 63–64, 181; II: 173, 439; III: 543 *see also* wildfire

— corrective, II: 443–444

— indigenous, I: 64; II: 463

— low-intensity surface, II: 287, 315, 318, 442, 454, 546

— natural, II: 444, 603

— prescribed *see* prescribed burning

fire model II: 68, 324

fire control, I: 67, 193; II: 292 *see also* fire suppression

fire frequency, II: 284, 289

fire gradient, II: 317

fire history, II: 298, 460, 485, 573

fire management, I: 194; II: 444–445

fire scar, II: 287, 298, 460

fire suppression, I: 42, 44, 67, 87, 191, 195; II: 43, 78, 175, 206, 266–267, 282, 289–290, 292–293, 317, 322, 453–455, 472, 475, 486, 524, 547, 604, 608, 653, 698; III: 197, 374

fish, I: 152

fish productivity, II: 610

fishery, I: 77

flexibility, I: 192; II: 5, 16, 28, 73; III: 158, 179, 600, 636

flood, I: 30, 100; II: 160

flood control, I: 86, 87, 187; II: 227, 319; III: 592

flood insurance, I: 200

floodplain forest, II: 168

flora, I: 55; II: 26–27

flying squirrel, Northern (*Glaucomys sabrinus*), II: 210, 455, 608

focus group, I: 117; III: 44, 52–53, 59, 77, 82, 235, 288

folk science, II: 457

Food and Agriculture Policy Research Institute (FAPRI), III: 682

food chain, II: 228; III: 407

food production, II: 73

food requirement, II: 70

Food Security Act, II: 531

food source, II: 26

food web, I: 34; II: 228, 259, 585; III: 404

Ford Foundation, III: 213

forest

— bristlecone pine, III: 269

— coastal lowland swamp, II: 73

— Douglas-fir, II: 302

— economic value, III: 385

— even-aged, II: 204, 291–292, 442

— higher elevation, II: 546

— late-successional, II: 572; III: 440

— lodgepole pine, II: 293

— longleaf pine, II: 528

— oak-hickory, II: 78

— old-growth, I: 194, 210; II: 8, 80, 264, 267, 549, 579, 607, 630; III: 33, 396, 440, 475, 523

— ponderosa pine, II: 575–576, 599, 603–606, 632; III: 269

— western hemlock, II: 599

— western hemlock-Douglas-fir, II: 289, 574

Forest and Rangelands Renewable Resources Planning Act (RPA), II: 6, 525, 694

forest area, II: 865

forest biomass, II: 530–531

forest certification, I: 161, 202; III: 245, 258–263

forest degradation, II: 604

forest dynamics model, II: 70

forest economics, United States, II: 695

Forest Ecosystem Management and Assessment Team (FEMAT), I: 193; II: 150–152, 607, 634–635; III: 255, 440–444

forest edge, I: 43

forest growth, II: 545

forest habitat, II: 537

forest health, I: 228; II: 545–548, 575, 635–636; III: 301, 348

forest history, United States, II: 693–695

Forest Incentive Program (FIP), III: 481

forest industry, I: 232; III: 146, 260

Forest Inventory and Analysis (FIA), II: 405; III: 656–658

forest legacy program, III: 481

forest management, II: 547, 571–579, 632; III: 281, 334, 348–349, 352–354

— sustainable, II: 418, 543; III: 259, 387, 475

forest plan monitoring, III: 653

forest planning model, II: 412

forest product, I: 155

forest regeneration, II: 608

forest stewardship, I: 200; III: 299

Forest Stewardship Council (FSC), II: 544; III: 259, 472–473

forest structure, II: 70

forest succession, II: 59, 71, 335

forestry, I: 77, 161; II: 557

formation, II: 364

fossil fuel, II: 523, 692–693; III: 333, 404

fox

— Sierra Nevada red, III: 522

— swift (*Vulpes velox*), II: 116

fragmentation, I: 135

framework

— conceptual, I: 160; II: 199; III: 274

— hierarchical, II: 397, 399

Freedom of Information Act (FOIA), I: 140; II: 12; III: 125, 147, 530

freshwater classification, II: 366

friction, cultural, I: 147

fringe

— anthropocentric, I: 80; II: 522

— biocentric, I: 81; II: 522

fringe group, I: 133

frog, leopard (*Rana pipiens*), II: 319

fuel, II: 315, 443, 454

fuel accumulation, II: 77, 285, 604

fuel reduction, III: 218

fuel removal, II: 605

fuelwood, I: 82

function

— ecological, I: 100; II: 626

— linear, III: 702

function indicator, II: 264–266

fund, I: 199, 230

funding, I: 139; III: 146

fungi, II: 49, 51, 59, 66

fungicide, II: 49

Gallatin National Forest, Idaho, II: 211; III: 395
game control, II: 648
Game Creek watershed, Alaska, II: 270–272
game population, II: 446
game species, II: 79, 561; III: 354
gamma diversity, II: 56
Gap analysis program (GAP), I: 58; II: 66, 185, 214, 337, 366, 375, 424; III: 609, 620, 647–650
gap dynamics, II: 587
gap formation, II: 300
garter snake, Mexican (*Thamnophis eques*), II: 81
gene bank, II: 532
gene flow, II: 12, 58, 88
gene pool, II: 338; III: 475
General Mining Law, I: 112; III: 10, 26–27
genetic loss, II: 72
genetics, I: 19, 23; II: 135
genotype, II: 72, 116
geoarchaeology, II: 460, 480
geographic information system (GIS), I: 17, 37, 51, 76, 90, 235, 238; II: 9, 12, 68, 70, 71, 121, 124, 140, 157, 184, 186, 196–197, 199–200, 262, 290–291, 332, 414, 487, 489, 573; III: 64, 143, 148, 270, 441, 556, 595, 609–610, 613–614, 619–626, 630–631, 641, 650, 653, 691, 694, 706–708
geology, I: 353; II: 165, 219, 224–227, 256
geomorphology, II: 165, 367, 480
Geospatial Data Clearinghouse, I: 639
Geothermal Steam Act, III: 26
germination, II: 300
ginseng, Chinese (*Panex ginseng*), III: 365
glacier, II: 225
Glacier National Park, Montana, II: 124
global assessment, I: 236
Global Biological Assessment, III: 334
Global Positioning System (GPS), I: 76; II: 508; III: 596, 610–611, 642, 681
gnatcatcher, I: 198; III: 464
goshawk, II: 299
Government, I: 200; II: 30; III: 480–484
Government Accounting Office (GAO), III: 560
Government control, III: 233
Government Land Office (GLO), II: 486
gradient
— altitudinal, I: 54
— climatic, II: 238
— environmental, I: 34; II: 45, 57–58, 68, 237, 356, 361, 408
— latitudinal, I: 54
— topographic, II: 283
grant, III: 117, 448, 484
grassland, II: 55, 444, 454, 684
grazing, I: 37, 97; II: 79, 205, 261, 268, 290, 473, 557, 576, 633; III: 378, 402 *see also* livestock grazing
Great Lakes Ecosystem, II: 8
Great Smoky Mountains National Park, II: 60, 275
Greater Yellowstone Ecosystem, I: 79, 99; II: 8, 125, 587
green accounting, I: 181; III: 346, 355–356
green certification, II: 533, 544–545
green credit card, I: 199; III: 473
green investment, III: 473

Green Mountain National Forest, III: 366
green product labeling, II: 544–545 *see also* ecolabeling
green production, II: 7
greenhouse gas, II: 698; III: 331, 333
grid, II: 110–111
grid cell, II: 418
grid line, II: 67
grid point, II: 67
grid system, III: 595
gross benefit, III: 350
gross domestic product (GDP), II: 544, 695–696; III: 331
gross primary productivity (GPP), II: 228
ground sloth, II: 446
ground water, II: 26, 168, 258, 367
group assessment, III: 105
group decision support system (GDSS), III: 699–700
group method, I: 120; III: 52
group stereotyping, I: 148
group thinking, III: 671
groupware, I: 244; III: 690–691, 697
grouse, sage, III: 605
growth
— economic, I: 188
— exponential, II: 93
growth model, ecological, II: 405
growth rate, II: 55, 79, 95, 96–99, 104–105, 107
guild, II: 13, 60, 139, 531
guild, birds, III: 575
guild structure, II: 374
gypsum soil, II: 166

habitat, I: 13; II: 128, 336; III: 625
— boundary, I: 29
— suitable, II: 100, 119–120
habitat area, I: 21, 22; II: 80
habitat attribute, II: 124
habitat capability, II: 142, 403
habitat condition, III: 441
habitat connectivity, I: 181
habitat conservation, I: 199; II: 195, 200–203; III: 621
Habitat Conservation Plan (HCP), I: 108, 110; III: 23, 146, 476
habitat corridor, I: 32; II: 533
habitat degradation, II: 48, 53, 58
habitat dynamics, II: 135, 141
habitat edge, II: 164
habitat evaluation, II: 120
Habitat Evaluation Procedure (HEP), II: 60, 62
habitat fragmentation, I: 20, 27, 43, 47; II: 8, 40, 53, 58, 69, 79, 88, 157, 170, 176, 210, 281, 303, 531, 562; III: 596, 692–697
habitat gradient, II: 119–120
habitat inventory, II: 142
habitat loss, I: 16, 20, 43; II: 143, 303, 318, 531, 686; III: 327
habitat modification, II: 442
habitat patch, I: 27
habitat quality, II: 120, 563; III: 396
habitat requirement, II: 60, 365
habitat restoration, II: 639

habitat selection, II: 119
habitat structure, II: 70, 79, 118–119, 122, 374
habitat suitability index (HSI), II: 62, 142
habitat threshold, II: 141
habitat transaction method (HTM), III: 478
habitat type, II: 119, 356, 369, 395, 405, 414–415, 420
habitat use, II: 338
habitat value, III: 622–624, 626
Haleakala National Park, Island of Maui, II: 564
hard-science model, III: 88
Hardy-Weinberg-expectation, II: 115
hare, snowshoe (*Lepus americanus*), II: 335
harvest, II: 450–451, 474
Hawaiian Islands, II: 563
Hawk Mountain Sanctuary, III: 372–374
hazard identification, III: 434–435
hazard reduction, I: 194
headwater zone, II: 260
heat exchange, II: 51, 233
heavy metals, II: 259
hemlock
— Eastern (*Tsuga canadensis*), II: 316
— Western (*Tsuga heterophylla*), II: 289, 442, 454, 574, 599, 607, 649
Henry Creek Grazing Allotment, III: 378–380
herbivore, I: 35; II: 50, 52, 71, 75, 77, 228
herbivory, II: 62, 648
heritage management, I: 71–76; II: 434, 493–516
heritage ranking system, II: 41
herptofauna, II: 81
heterogeneity, I: 15; II: 41
— environmental, II: 56–57, 62–63
— spatial, II: 114, 585
heterozygosity, II: 140, 338
Hiawatha National Forest, III: 169
hierarchy, II: 25, 32, 345, 368
— biological, II: 219, 221, 333, 345
— human, III: 537
— organizational, I: 126
— scale-time, II: 344
— sociocultural, I: 5; II: 32, 345
— space-time, II: 315, 333
hierarchy process, analytical (AHP), I: 243; III: 691, 697, 703–705, 716
hierarchy theory, II: 333, 358; III: 537
Hispanic Association of Colleges and Universities (HACU), III: 201
Hispanic Radio Network, III: 219
Hispanic Serving Institution (HSI), III: 201
historical science, II: 493–516
history
— environmental, III: 189, 204–205
— oral, II: 483; III: 51, 235
Holdridge life zone system, II: 238–239; III: 228
holism, II: 9, 203, 552–553, 584, 622; III: 55, 88, 183, 281, 297, 300–303, 664
Holocene, II: 479
home range, II: 120, 128, 144, 146, 149
homeostasis, II: 223, 235

Homestead Act, III: 10
honey mesquite (*Prosopis glandulosa*), II: 452
honeysuckle, Japanese, II: 292, 304, 631, 637
Hoosier National Forest, Indiana, I: 75; II: 423, 507
horse, wild (*Equus* spp.), II: 446
horticulture, II: 451–453
household, III: 233, 250
housing development, I: 198; II: 73, 455, 697; III: 175, 464, 620
Hubbard Brook Ecosystem Study, II: 8, 22, 166; III: 287
human behavior, II: 23, 29; III: 135, 230, 280
human capital, II: 27, 544; III: 233
human carrying capacity, II: 677, 689–692
human choice, II: 690–691
human culture, II: 522
human ecology, II: 21, 23–24; III: 229, 285
human ecosystem, II: 23–32, 494
human health, II: 525–526
human history, II: 534, 558
human impact, I: 15, 43; II: 22, 173–179, 303, 471–492, 496; III: 622–624 *see also* human role
human population, I: 169; II: 439
human population growth, I: 82, 170, 171; II: 22, 27, 39, 528, 534, 680–701
human population growth rate, II: 681; III: 328, 330
human role, I: 63–70; II: 433, 485; III: 44 *see also* human impact
Humboldt National Forest, III: 269
humic acids, II: 51
humidity, II: 57, 338
humidity provinces, II: 238–239
humility principle, II: 44
humpback chub (*Gila cypha*), II: 319
humus, II: 230
hunter-gatherer, II: 439, 447–450, 502
hunting, II: 175, 445–450
— historical, II: 499–500
hunting technology, II: 447–450
hurricane, II: 236
hybridization, II: 116
hydroelectricity, II: 26
hydrologic unit, II: 368
hydrology, I: 44; II: 73
hypermedia, III: 290, 698–699
hypothesis testing, I: 116, 219; III: 604
hysteresis, II: 223

ibis, white faced, II: 654
ice age, II: 225
Idaho Panhandle National Forest, II: 578
immigration, I: 172; II: 60
impact
— cumulative, I: 48
— potential local, I: 174
implementation monitoring, III: 31, 67, 589–590
inbreeding depression, II: 48, 58, 61, 88, 98, 115–116, 117–118, 144, 149, 334, 338
incentive, I: 133; II: 11, 14; III: 322, 422, 447, 481–482, 527
— coinciding, III: 354
— conflicting, III: 354

— economic, I: 181, 197; II: 11; III: 347, 379
— Government, I: 200; III: 324, 389
— perverse, II: 540; III: 353, 465
— positive III: 353–354
income, I: 179; III: 351
income range, II: 32
income tax, I: 198; III: 469
indicator, I: 85, 223, 225, 226, 227; II: 42, 256, 265, 274, 592–597; III: 564, 575–577, 592–593
— biophysical, II: 32
— cause and effect, I: 92
— cost-effectiveness, I: 92
— diagnostic, I: 227; III: 575–576
— ecological, I: 91
— ecosystem condition, I: 87, 90, 92, 188; II: 264, 592–597
— environmental change, II: 53
— predictive, I: 227; III: 575
— social, I: 154; II: 32, 544; III: 232
— socioeconomic, III: 65
— usability, I: 92
indicator assessment, I: 227; II: 583, 594; III: 575
indicator selection, II: 595–596
indicator species, II: 139, 364, 602 see also keystone species
indicator variable, I: 220; III: 543
Industrial Revolution, II: 505, 690; III: 86
industrialization, I: 82, 505; II: 480; III: 251
industry, I: 5; II: 29; III: 530
INFISH strategy, III: 33
information, I: 6, 169, 209, 211–212, 228, 232, 234–235; II: 12, 27; III: 448
— acquiring, I: 136; III: 516
— baseline, I: 218
— demographic, I: 117; III: 45
— dose-response, I: 92; III: 436, 438
— existing, III: 48–49
extrapolation, II: 350
— feedback loops, I: 7
— integrated scientific, III: 14, 633
— large-scale, II: 350
— monitoring, I: 218
public, I: 124
— quality, I: 17
— quantitative, I: 57
— site-specific, III: 556
information engineering (IE), III: 633
information era, III: 148, 630
information failure, III: 388
information flow, I: 229
information framework, I: 53
information level, II: 412
information management, I: 217; III: 556–557, 629–660
information need, III: 605
information needs assessment (INA), I: 236–237; III: 631–633, 646, 652
information planning, I: 235–236; III: 631
information process, III: 310
information scale, II: 348–351
information sharing, I: 3, 129, 130, 132, 133, 236; III: 6, 97, 100, 106–109, 113, 147–148, 158–159, 170, 439, 690

information source, I: 174; II: 458; III: 568
information technology, I: 231; II: 9; III: 335
information theme, III: 557
information tool, III: 336–340
information transfer, I: 233; II: 168; III: 607, 609
inheritance tax, III: 471
innovation, I: 161, 173; III: 287
insect, II: 61, 66, 286
— aquatic, II: 54
— outbreak, I: 43; II: 293; III: 334
insecticide, II: 49
institution
— educational, II: 15
— political, II: 692
— social, II: 21, 24, 25, 28–31
instream structure, II: 642
integrated environmental management (IEM), III: 186, 302–303, 312–315
integrated resource inventory (IRI), II: 576
Inter-University Consortium for Political and Social Research (ICPSR), III: 49
interaction, I: 211
— animal–ecosystem, II: 180
— bear–human, III: 111–112, 125–126, 237
— biological, I: 15
— cooperative, III: 134
— ecologic–economic, I: 188; III: 404
— economic, I: 177–183; III: 345–358
— human–ecosystem, III: 5
— human–environment, I: 64, 67, 73; II: 459, 476; III: 190, 203, 222, 228
— human–plant, II: 458
— human–wildland, I: 98
— land–water, II: 168–171
— regional, III: 349–352
interagency agreement, III: 113, 116, 177
interagency consultation, III: 21
interagency cooperation, III: 164, 554
Interagency Ecosystem Management Task Force, II: 4
interagency management, III: 159
interagency network, II: 565
Interagency Scientific Committee (ISC), II: 125
Interagency Steering Group (ISG), III: 164–165
interagency team, I: 216
interagency vegetation standard, III: 646–647
interdisciplinarity, I: 7, 69, 97; II: 489, 542, 625–627; III: 305–306, 557
interdisciplinary team (IDT), III: 305, 422, 459, 520, 558, 590, 690
interest
— economic, III: 262
— mutual, III: 135
— public, III: 365
interest focus, III: 89, 102, 118, 202, 702
interest group, II: 699
interest scale, III: 61
interface
— atmospheric–terrestrial, II: 241
— biotic, II: 241

— ecological, II: 240
— functional, II: 237–242
— plant–soil, II: 241
— terrestrial–hydrologic, II: 241
Intergovernmental Panel on Climate Change (IPCC), III: 333
Interior Columbia Basin Ecosystem Management Project (ICBEMP), I: 119, 181, 216; II: 266–269, 336, 570, 623, 634–635; III: 33, 70, 204, 367–372, 650–652
Interior Columbia River Basin Project (ICRB), III: 536, 540–541
intermediate disturbance hypothesis (IDH), II: 64, 301
Internal Revenue Service (IRS), III: 466
International Biological Program, III: 615
International Electrotechnical Commission (IEC), III: 615
International Standards Organization (ISO), III: 615, 640
Internet, I: 174, 235; II: 12; III: 49, 64, 291, 616, 690, 698–699
intervention
— biological, II: 637–640, 648–649, 653–656
— Government, III: 388
— human, II: 659
intervention management, III: 518, 521
interview, III: 46, 51, 68, 77, 223, 235, 268
Inuit, II: 449
inventory, I: 231–239; II: 262; III: 146, 291, 520, 603–628, 631, 643–646
— ecological, II: 401, 404; III: 654–655
— integrated, II: 208; III: 652–653
iron wood (Olneya tesota), II: 53
irrigation, I: 74; II: 74–75, 226, 655
island biography, II: 40, 163, 200
island ecosystem, II: 539
isolation, II: 58, 69
isopleth, II: 238

jay
— Eastern blue, II: 81
— Florida scrub, II: 113
joint production, I: 212; III: 362–363, 370, 515
judicial deference principle, III: 34
juniper (Juniper spp.), II: 454
justice, II: 28–29
juxtaposition, II: 162, 332

kelp, II: 590
— giant, II: 52
Keystone Dialogue, I: 2
keystone resource, II: 80
keystone species, I: 20, 31, 34, 35, 100; II: 22, 52, 80, 139, 179–181, 244, 258, 563, 588, 590; III: 283
Kisatchie National Park, Louisiana, I: 28; II: 209
kite, Everglades (snail), II: 113
knapweed
— diffuse, II: 638
— featherhead, II: 638
— Russian, II: 53
— spotted (Centaurea maculosa), II: 52, 181, 488, 637–638
knowledge, I: 136, 207, 208, 211, 232; II: 10, 27, 29
— collective, I: 213

— cross-translation, III: 507
— cultural, I: 147
— ecological, II: 436, 456
— empirical, III: 288
— environmental, II: 456; III: 240
— experienced-based, III: 517
— historical, II: 624
— indigenous, II: 456; III: 52, 262
— interconnectedness, III: 699
— local, I: 145, 174; II: 461, 465; III: 185, 196–197, 204, 340
— professional, III: 249
— public, III: 708
— scientific, II: 7–8, 495; III: 262, 418
— site-specific, II: 623
— technical, II: 505
— traditional, II: 439, 456; III: 189, 227
knowledge-based systems, II: 449; III: 688–689, 691–692, 708
Knutson–Vandenberg Act, III: 529
Kootenai National Forest, Montana, II: 484–485
kudzu vine, I: 16; II: 52, 206, 292, 304, 637
Kyoto Protocol, III: 331

labor, I: 186; II: 27; III: 351
labor market, I: 155
lake classification, II: 367, 398
lake type, II: 168
land, II: 26
Land and Resource Management Plan (LRMP), III: 18, 30–31
Land and Water Conservation Fund Act, II: 565
land classification, III: 146
land cover, II: 375, 395
land degradation, II: 679
land donation, I: 198; III: 467, 469
land ethic, II: 9
land exchange, I: 198; II: 484; III: 70, 468
land grant, III: 200
land management, II: 14, 526, 541; III: 32, 269
land management, federal, I: 5, 80, 83; II: 14
land management, public, III: 29, 103
land manager, I: 109
land manager, federal, II: 15
land manager, public, III: 4
Land Resource Region (LRR), II: 421–422
land survey, I: 67
land tenure, II: 32
land trust, I: 198; III: 468
Land Trust Alliance, III: 466
land type, II: 406, 414
land type association (LTA), II: 209, 399, 402, 407, 412–413; III: 170, 556, 652, 655
land unit, ecological, II: 336
Land Unsuitable Mining Program (LUMP), II: 9
land use, I: 29, 43, 99, 170; II: 621, 656–663
— change, II: 32, 174–176, 688–690; III: 621, 624
— extensification, II: 684, 688, 700
— intensification, II: 684, 688, 700
— history, I: 72, 74, 76; II: 458, 683–684

— planning, I: 17, 202; II: 157, 356; III: 27–28
— traditional, III: 196
— shift, I: 67, 169–175
— trade-off, II: 437
— transformation, II: 502
Land Use History of North America (LUHNA), II: 466
landform pattern, II: 506
landform unit, II: 406
landowner incentive, III: 352
landowner type, II: 177
landownership, II: 53, 626; III: 364, 693
— private, II: 540; III: 65, 143
— public, I: 77; III: 354–355
Landsat Multi-Spectral Scanner (MSS), III: 642–643
Landsat satellite, I: 235
Landsat Thematic Mapper (TM), III: 642–643, 648, 653, 655
landscape, I: 27, 29, 36, 47, 209; II: 162, 205, 222; III: 161, 309
— cultural, I: 73–74; II: 489–490, 499–503, 508
— heterogeneity, I: 29; II: 197
— historic, II: 471
— industrial, II: 505
landscape analysis program, FRAGSTATS, II: 210
landscape approach, II: 195, 198–200
landscape assessment, I: 58; II: 184, 210, 559; III: 164
landscape boundary, I: 29, 48; II: 162, 182–183
landscape change, I: 41; II: 60, 184
landscape data, II: 187
landscape diversity, II: 165–173, 173–184, 195–218
landscape ecology, I: 27–28; II: 4, 158, 282, 410, 418, 559; III: 537, 615
landscape hierarchy, II: 158
landscape model, II: 71–72, 335; III: 679
landscape mosaic, I: 224; II: 281, 290–294
landscape painting, II: 483
landscape pattern, I: 32, 41; II: 42, 161–165, 197, 281, 290
landscape perspective, II: 623–624
landscape planning, I: 58; II: 331; III: 30
landscape scale, II: 41, 67, 265, 295
landscape structure, II: 53, 60, 162; III: 575
landscape survey, II: 360
landscape value, II: 473
landslide, I: 41, 42, 100; II: 168, 172, 225
language extinction, III: 193
larch, western, II: 635
larkspur, tall, II: 78
late serial forest (LSF), II: 574
law, I: 6, 74, 107; II: 8, 28–29; III: 4, 677
— changes, II: 14
— federal, II: 196
— natural, II: 43
— physical, I: 188
leadership, II: 9; III: 101, 272
learning, I: 207, 208, 209; II: 8, 29; III: 6, 515
— adaptive, II: 436
— collaborative, I: 131; III: 103–105, 588
— effective, III: 506, 698
— expert-based, I: 210
— just-in-time, III: 698
— long-term, III: 516

— mutual, III: 108, 307, 554
— objective, I: 212; III: 517
— organizational, III: 245, 248–249, 419
— parallel, III: 505, 512, 518–519, 521, 529
— reactive, III: 506
— sequential, III: 512, 518–520, 529
— short-term, III: 516
— social, III: 248, 252, 529, 552, 554
learning organization, I: 126, 161, 163; III: 93
learning strategy, III: 508, 516
leasehold property right, III: 26–27
legacy
— biological, I: 41; II: 42, 287, 290, 656
— ecological, II: 433
legal framework, I: 108; III: 10, 14–28
legal requirement, I: 110
legal system, II: 29
legislation, I: 107–113, 115; III: 36
— federal, I: 109
— organic, III: 4
— site-specific, III: 36
legitimacy, III: 142
leisure, II: 29
Leslie matrix projection model, II: 93–94
levee, I: 86
liability, I: 188; III: 399
lichen, II: 59
life cycle, II: 170
life cycle analysis, II: 545
life cycle management (LCM), III: 633–635
life cycle planning, I: 236; III: 633–635
life form, I: 35; II: 240, 364
life span, II: 335
life support system, II: 589; III: 405
life zone, I: 53; II: 238, 247, 365
likelihood, III: 418, 433
likelihood classification, III: 612
Lincoln National Forest, New Mexico, II: 633
linear thinking, I: 131
lion (Panthera leo), II: 446
lithology, II: 571
litigation, I: 107, 111, 140, 216; II: 31, 125; III: 19, 32, 178
Little Ice Age, II: 684
livestock, I: 37, 178, 202; II: 26, 50, 268, 473
livestock grazing, II: 53, 175, 260, 299, 407, 626, 644–645, 646, 654; III: 17 see also grazing
livestock management, II: 565–566
lizard, Coachella Valley fringe-toed, III: 142, 580
local area network (LAN), III: 650
logging, I: 42, 198; II: 493; III: 349 see also clearcutting
Long Term Ecological Research (LTER), II: 166
longleaf pine-wiregrass ecosystem, II: 586, 588–590, 630
loosestrife, purple, I: 16, 31; II: 638, 656
Los Angeles, II: 213–214
Los Angeles Conservation Corps (LACC), III: 220
Lotka–Volterra competition model, II: 69
lousewort, Furbish's (Pedicularus furbishiae), II: 110
lumber, I: 82; II: 26
— consumption, II: 538

lynx, Canadian (*Felix lynx*), II: 335

machine model thinking, III: 85–87, 90–92
maize, II: 501
Major Land Resource Area (MLRA), II: 421–422, 424
Malapai Borderlands Group, I: 137
Malthus–Condorcet–Mill model, II: 689–690
mammoth (*Mammuthus* spp.), II: 446
management, I: 211, 245; III: 30, 509–510
— adaptive *see* adaptive management
— alternative, III: 291, 417, 559, 678–679
— bioregional, I: 2
— collaborative, I: 127; III: 134
— comparative, III: 521–522
— ecological, I: 4
— experimental, III: 521–522, 524, 526
— holistic, II: 357
— multiple-use, I: 78, 79, 81, 124, 126; II: 4, 45, 208–210,
   400, 517, 519–521, 538, 558; III: 312
— of processes, I: 139; III: 141, 145, 157, 163
— organizational, III: 149
— passive, II: 634
— plan-level, I: 110; III: 30–31
— site-specific, III: 30, 32
— sustainable, I: 33, 127
management action, I: 17, 225
management area (MA), II: 412
management effectiveness, III: 564, 576
management era, III: 10
management objective, I: 233, 243; III: 608, 670
management plan, III: 161
management practice, I: 36; II: 621, 462–463
management scenario, II: 404, 411
management science, I: 242
management strategy, III: 598–600
management success, III: 577–579
management threshold, III: 608
management trade-off, II: 400, 589
mandate, legal, III: 161
mangrove, II: 236, 367, 652
map, I: 53, 54, 217, 225; II: 43, 66; III: 566, 595, 613
— cognitive, III: 258, 261
— digital, II: 68
— thematic, III: 611
map boundary, II: 397
map unit, ecological, I: 55
maple, sugar (*Acer saccharum*), II: 316, 442, 454
mapping, I: 57, 238; II: 331, 370, 395, 402–403; III: 645
mapping scale, II: 411
Mark Twain National Forest, III: 653–654
market
— borax, III: 195
— global, I: 180; II: 503, 686
market economy, III: 321
market emulation, I: 188; III: 400
market force, III: 365
market hunting, II: 523, 529
market price, III: 349
market value, II: 535–536; III: 359, 394

Markov model, III: 438
marsh, II: 59, 319, 599, 610–611
marsh elder (*Iva frutescens*), II: 610
marsh-estuarine system, II: 610
marsh restoration, II: 610–611
marten, pine, II: 299
mass wasting, II: 226
mastodon (*Mammut americanum*), II: 446
material cycle, II: 219, 227–233, 256
material exchange, II: 584
material flow, I: 34; III: 282
material transfer, II: 260
material well-being, II: 691
matrix, I: 29; II: 162
matrix analysis, I: 188
matrix model, II: 69, 93–94
matrix selection method, II: 106
matrix technique, III: 672, 674, 679–680
measure, validity, I: 153
measurement analysis, II: 45
measurement endpoint, III: 418, 433
measurement technique, III: 288
media, I: 169, 173, 192; II: 483
media accounts, I: 174
mediator, III: 119, 121
medicine, I: 201; II: 28
— holistic, III: 301–302
— natural, II: 536
medicine man, II: 451
megafauna extinction, II: 446
Memorandum of Agreement (MOA), III: 34, 128, 159
Memorandum of Understanding (MOU), II: 384, 399, 422;
   III: 81, 34, 116, 159, 161, 164–165, 168, 201, 223, 270–271
memory, institutional, III: 526
metadata, II: 187; III: 636, 640
metapopulation, II: 12, 58, 69, 91, 108–109, 128, 145, 214, 338
meteorite, II: 227
method
— archival, I: 229
— command-and-control, III: 423
— expert-intuitive, III: 668
— fault tree analysis, III: 435, 438
— garbage can, III: 667, 674
— hypothetical-direct, III: 391–392
— muddling through, III: 668
— normative-rational, III: 667
— paleobotanical, II: 476–480
— programmatic, III: 22
— qualitative, III: 74
— retrospective, II: 604; III: 525–526, 520, 530
— scale of analysis, II: 139, 347–348, 475
— scientific, II: 15
— statistical, I: 211; II: 68; III: 53, 66, 543, 594–595, 679
— strong inference, II: 15
— tabular, II: 362
— visual, III: 614
microclimate, II: 51, 170, 205, 236, 293
microcosm, I: 179
microenvironment, II: 301

micromanagement, III: 671
microsite heterogeneity, II: 182
migration, I: 170, 172; II: 27, 69, 447
— global, II: 505–506
— internal, III: 328
— of species, II: 182
migration capacity, II: 170
mine land, I: 100–101
mineral, stability, II: 165
mineral assessment, II: 569–570
mineral cycling, II: 166
Mineral Leasing Act, III: 11, 26
mineral resource map, II: 571
mineralization, II: 231
mining, II: 226, 275, 493, 504–506, 557, 570, 657, 658, 660; III: 194, 349
mining industry, II: 504, 506
mite, tracheal, II: 56
mitigation, II: 275, 648; III: 20, 123, 425, 433
mitigation banking, III: 478
mitigation measure, III: 417
model, I: 17, 241; II: 72; III: 568–570
— allocation, II: 335
— Beverton-Holt, II: 97–98, 101, 113
— biophysical, II: 68–69
— conceptual, I: 243; II: 274
— core-satellite, II: 109
— deductive, I: 116; III: 45
— demographic, II: 112–115
— derivative, II: 24
— deterministic, III: 567
— discrete logistics, II: 97
— dynamic equilibrium, II: 64–65, 77
— dynamic gap, II: 92
— ecological, I: 223, 225, 226, 227; III: 566–570, 575
— exponential growth, II: 94–95
— fate, III: 438
— FIBER, II: 405–406
— fisheries-stream reach, III: 679
— four-phase, II: 300
— genetic, II: 115–118
— geochemical, II: 660
— habitat-relations, II: 120–121
— hydrologic, II: 68
— inductive, III: 45
— LANDSCAPE, II: 112
— Levin's metapopulation, II: 109
— limitation, II: 145
— mainland-island, II: 109
— map-based, II: 121–122
— mathematical, III: 708
— Maynard-Smith and Slatkin, II: 99
— mechanistic, I: 123; II: 99–100
— mental, III: 673
— of processes, II: 404
— organic, I: 123, 126; III: 6, 85–87, 90–92
— patch extinction/colonization, II: 110
— PATREC, II: 142
— POET, II: 24, 27
— predictive, II: 138
— qualitative, II: 688–689
— quantitative, II: 689–690
— Ricker, II: 97–98, 113
— simulation see simulation model
— soil moisture, II: 68
— spatial, II: 92, 110–112, 123
— stand-level growth, III: 679
— state-transition, III: 574
— statistical, III: 438
— stochastic, II: 102–107; III: 438
— structured, II: 106
— transport, III: 438
model process, I: 211
model uncertainty, III: 414
model use II: 68–72
model validation, II: 122
monetization, III: 389
monitoring, I: 90, 98, 99, 207, 209, 210, 215, 218, 223–230, 231, 236, 245; II: 274, 322–324, 400, 416–418, 609, 625; III: 20, 114, 127, 138, 291, 435, 453, 523–525, 564, 696
— ecological, I: 224
— effectiveness, I: 229
— indicator, II: 323; III: 566–567
— institutionalization, III: 598
— key parameter, II: 323
— long-term, I: 211; II: 255
— multiple scale, II: 255
— objective, I: 226, 228; III: 589
— participatory, I: 131, 225
— validation, I: 229
monitoring data, I: 229
monitoring design, III: 586–587
monitoring plan, I: 225
monitoring program, I: 227; III: 421, 546, 586
monitoring purpose, I: 228
monitoring requirement, III: 31
monitoring system, II: 248
monitoring techniques, II: 323
Monongahela National Forest, Virginia, II: 210–211, 525–526
Monte Alege Paleoindian, II: 441
Montreal Process, II: 543
moose (Alces alces), I: 157; II: 76, 180, 446; III: 238
moral hazard, III: 388, 400
morphology, II: 46
mortality, II: 64, 96, 99, 236
mortality rate, II: 40
moth, Gypsy, I: 16; II: 289, 304, 637
Mount Hood National Forest, III: 521
Multi-Resolution Land Characterization (MRLC), II: 375; III: 642, 648
multiple use, I: 77; III: 212, 471
Multiple-Use Sustained Yield Act (MUSY), I: 108; II: 5, 435, 525, 694; III: 11, 18, 348, 665
multivariate analysis of variance (MANOVA), III: 701
multivariate method, I: 55; II: 361, 399
murrelet, marbled (Brachyrampus marmoratus), I: 232; II: 264, 607, 620; III: 142, 440

mussel, zebra, I: 16, 22; II: 137, 410
mutualism, I: 15; II: 13, 47, 56, 62, 70, 101–102, 588; III: 544
mycorrhiza, II: 13, 62

Nantahala National Forest, North Carolina, II: 324
National Acid Precipitation Assessment Program (NAPAP), III: 334
National Aerial Photography Program (NAPP), III: 654
National Aeronautics and Space Administration (NASA), II: 187
National Agricultural Library Catalogue (AGRICOLA), III: 49
National Biological Information Infrastructure (NBII), II: 186
National Biological Service (NBS), II: 424; III: 169, 648
National Environmental Protection (Policy) Act (NEPA), I: 107–110, 146, 212; II: 5, 9, 196, 200, 203, 414, 525, 622; III: 3, 9, 11, 13, 15, 20–22, 78, 99, 190, 204, 210, 272, 305, 368, 440, 451, 515, 517, 536, 671, 677, 690
National Forest Management Act (NFMA), I: 108–109; II: 5–6, 46, 136–138, 146, 151, 196, 200, 208, 346, 414, 525–527, 541, 694; III: 4, 9, 11, 13–14, 18–19, 150, 210, 348, 368, 440, 450, 665, 677
National Forest System (NFS), II: 527, 542; III: 656
National Geographic magazine, I: 72; II: 508
National Heritage Area, III: 171
National Hierarchical Framework of Ecological Units (NHFEU), II: 184, 208, 398–399, 405, 475; III: 161, 170, 542
National Historic Preservation Act (NHPA), I: 74, 146; II: 460, 477, 483, 495, 509; III: 190, 211
National Institute for Standards and Technology (NIST), III: 640
National Marine Fisheries Service (NMFS), I: 110; III: 23–24
National Oceanographic and Atmospheric Administration (NOAA), III: 648
national park, III: 15–16
National Park Service (NPS), I: 109; II: 319, 524; III: 11, 69, 169, 195, 591
National Park Service Organic Act, I: 108; III: 11
National Register of Historic Places, I: 72, 75; II: 435, 477, 484, 507
National Science Foundation (NSF), II: 187
National Spatial Data Infrastructure (NSDI), I: 238–239; II: 185; III: 637–639, 644
National Survey of Fishing, Hunting and Wildlife Associated Recreation (NSFHWAR), III: 64
National Survey on Recreation and the Environment (NSRE), III: 64
National Trails System Act, II: 525
National Water Quality Assessment (NAWQA), I: 58; II: 382, 423
National Wilderness System, II: 6
National Wildlife Refuge System Administration Act, I: 108; III: 17
Native Alaskan, III: 238–240
Native American, I: 63–70, 71, 146, 147; II: 174, 259, 319, 433–434, 472, 487, 603; III: 4, 113, 177, 190–196, 203, 539
Native American Graves Protection and Repatriation Act (NAGPRA), III: 190, 211

Native American Languages Act, III: 193
Native American
— environmental impact of, I: 65; II: 439–470
— population decline of, II: 441
Native American culture, I: 66, 73
Native American elder, II: 459, 484
Native American interests, III: 81
Native American management practice, II: 441–442, 456
natural capital, I: 185; II: 544; III: 355, 364
natural ecosystem manager see keystone species
natural gas, II: 26
Natural Heritage Program (NHP), II: 67, 203, 423
natural resource, II: 24–27
— primary, III: 591
— secondary, III: 591
natural resource assessment, I: 215; III: 553
natural resource consumption, II: 536; III: 328–330
natural resource economy, III: 234
natural resource inventory, III: 652–653
natural resource management, I: 1, 13, 58, 77–78; II: 517–556 see also ecosystem management
natural resource manager, I: 124–126
natural resource planning, I: 58
natural resource production, I: 77–84; II: 22, 528, 535
natural resource use, I: 147, 171, 173; III: 331–333
Natural Resources Conservation Service (NRCS), II: 424, 524–525
natural science, II: 679
natural selection, II: 48
naturalness, II: 602–603
nearctic zone, II: 402
negotiation, III: 97, 121–122, 150, 302, 399
neighborhood diversity, III: 623–624
nematode, II: 231
neotropical migratory birds (NTMB), II: 415–416
nested forest hierarchy, II: 197–198
nesting, II: 80, 654
net ecosystem productivity (NEP), II: 228
net present value (NPV), II: 574; III: 376
net primary production (NPP), II: 71, 75–77, 228, 243, 423, 605–606 see also primary production
nettle (Urtica dioica), II: 453
network, I: 135, 218, 238; II: 162; III: 148, 194, 257, 266
network model, II: 22
networking, III: 65, 140
New England, forest land, I: 83
New Forestry movement, II: 634
Newtonian thinking, III: 92
niche, II: 56, 62–63, 333, 504–505, 608; III: 184, 193
nitrification, II: 232
nitrogen, II: 50, 68
nitrogen cycle, II: 171, 231
nitrogen deposition, II: 233
nitrogen fertilization, II: 233
nitrogen flux, II: 171
nitrogen input, II: 76
nitrogen loss, II: 232
noise, I: 51
non-cooperation model, III: 135

non-governmental organization (NGO), III: 265, 467
non-impairment standard, III: 15–17
non-industrial private forest (NIPF), III: 481
non-intervention, II: 661
non-market good, III: 352
non-market resources, II: 536–537
non-profit organization, III: 174
norm, II: 25, 29, 31–32
North American Landscape Characterization (NALC), III: 642
Northside Ecosystem Project, III: 374–378
Northwest Forest Plan (NWFP), II: 10, 146–148, 150–152, 270, 573–574; III: 32–33, 513, 524
nutrient, II: 26
nutrient availability, I: 15; II: 46, 173
nutrient cycle, II: 49, 68, 243, 256, 605; III: 282, 570
nutrient depletion, II: 76
nutrient recycling, II: 228
nutrient transfer, II: 260
nutrient transformation rate, I: 90

oak (Quercus spp.), II: 335, 454
Oak Ridge, Tennessee, II: 63
oak, chestnut (Quercus prinus), II: 51
objective analysis, III: 669
objective refinement, III: 677–678, 690, 693
observation, II: 287; III: 77
— direct, I: 234; III: 610, 616
— empirical, III: 285
— indirect, III: 391–392
— participant, III: 51
Ochoco National Forest, II: 578
Office of General Council (OGC), III: 560
Office of Management and Budget (OMB), II: 14, 424; III: 67
Oil Pollution Act (OPA), III: 24–25
open system, II: 584
opinion polling, III: 45, 49
opinion, public, I: 115
opportunity, I: 192, 194; III: 157, 432
opportunity cost, II: 591; III: 322, 360, 400, 449
Opportunity Los Angeles (OLA), III: 219
opposition, public, III: 144, 149
optimizing tools, I: 244
ordination, TWINSPAN, II: 362
organic act, I: 108; III: 15
Organic Administration Act, III: 11
organic matter, II: 230–231, 234
organic model thinking, I: 123, 126; III: 6, 85–87, 90–92
organic web, II: 24
organization, II: 24, 27–28, 358–359; III: 87, 287, 309, 456, 635, 673
Ottawa National Forest, II: 412
Ouachita National Forest, Arkansas, I: 210; II: 322, 624, 639; III: 74, 528
output, I: 178
non-market, III: 363
output product, I: 237
output quantity, III: 89

over-population, II: 27
overgrazing, II: 75, 506, 567, 604
overharvest, I: 16
owl
— California spotted (Strix occidentalis), II: 455
— flammulated (Otus flammeolus), II: 562
— great grey (Strix nebulosa), II: 455
— Northern spotted (Strix occidentalis caurina), I: 24, 111, 198, 232; II: 5–6, 10, 79–80, 96, 99–100, 106–107, 113, 119–123, 125, 141, 144, 146–148, 264, 455, 527, 607–608, 620, 629; III: 19, 32, 247, 350, 440, 464, 513, 522, 590, 646
ownership, I: 217; II: 8, 344
— pattern, II: 26
— private, I: 187
— public, III: 272
ownership objective, II: 177–178
oxygen, II: 64, 227
ozone, II: 53
ozone-depleting chemical, I: 189

PACFISH strategy, III: 33
Pacific Northwest, Coastal Ecosystem, I: 37
Pacific Northwest, Western Cascades, II: 264–266
Pacific Northwest Forest, I: 193, 232; II: 606–610
packrat middens, II: 478–479, 487
Paleoindians, II: 440
paleolimnology, II: 460
palo verde (Cercidium), II: 53
palynology, II: 460
panic grass (Panicum sonoran), II: 452
panther, Florida, II: 61, 72, 164
paper consumption, I: 171
paper functioning condition (PFC), II: 566
paper mill, III: 334, 367
paradigm, I: 207
— declining population, II: 88
— natural resource management, I: 78
— prevailing management, III: 403, 408
— small population, II: 89
paradigm shift, III: 5, 86, 664–665, 672
parasitism, I: 15
participatory action research (PAR), I: 117, 209; III: 6, 47, 78, 186, 267
participatory rural appraisal (PRA), I: 131; III: 52, 203
Partners in Flight (PIF), II: 415
partnership, I: 135, 210, 213, 217; II: 4, 14, 422–425; III: 97, 100, 115–118, 155, 599
— citizen–manager–scientist, I: 209; III: 506, 525
— collaborative, I: 98, 161; II: 7, 42
— community-based, III: 253, 479
— process design, III: 116–117
— public, II: 435
— public–private, I: 79, 81, 199–200; II: 519, 553, 629; III: 132, 464–465, 476–479
— voluntary, II: 213
partnership framework, II: 565
patch, I: 29; II: 162
patch boundary, II: 291

patch effect, II: 657
patch model, II: 107–108, 110
patch type, I: 41; III: 580
pathogen, II: 80, 286; III: 334
pathway
— evolutionary, I: 153
— successional, II: 242, 434
pattern
— coarse-scale, II: 282
— hump-backed, II: 64
— temporal, II: 58–60
pattern recognition, II: 358
Payette National Forest, II: 207
peatland, II: 367
peer review, I: 238, 243, 245; III: 502, 527, 616, 648, 678
performance bond, III: 400
performance criteria, II: 542–543
performance evaluation system, II: 14
performance indicator, III: 451
persistence, II: 105–106, 114–115, 120, 141, 151, 295, 456; III: 287
pest, II: 55, 75, 80, 657
pest management, integrated, III: 485
pest outbreak, II: 54–56, 78, 264, 288–290
pesticide, II: 6, 53, 525, 530
pharmacopoeia, II: 26
phosphorus availability, II: 284
photography, I: 238; II: 461
— aerial, I: 17, 30, 55; II: 67, 69, 173, 325, 344; III: 268, 580, 594, 620, 643, 653
— historic, I: 67; II: 488
photosynthesis, II: 71, 224, 227, 231, 233; III: 577
physiography, I: 55
phytolith analysis, II: 477
phytoplankton, III: 573
Pickett Act, III: 11
pickleweed, III: 194
pig, II: 563–564
pigeon, passenger (*Ectopistes migratorius*), I: 21–22; II: 61, 80, 175, 257, 440; III: 464
pike, blue (*Stizostedion vitreum glaucum*) II: 410
Pike National Forest, III: 451
pine
— jack (*Pinus banksiana*), II: 317
— loblolly (*Pinus taeda*), I: 31, 192; II: 315
— lodgepole (*Pinus contorta*), I: 31, 42; II: 173, 293, 454
— longleaf (*Pinus palustris*), I: 28; II: 42, 209, 315, 588; III: 464, 528, 586, 588–590, 630
— ponderosa (*Pinus ponderosa*), I: 42, 97; II: 173, 221, 267, 318, 575–577, 579, 599, 603, 630, 632; III: 269, 374, 543
— red (*Pinus resinosa*), II: 317
— shortleaf (*Pinus echinata*), II: 315
— sugar, II: 451
— Torrey (*Pinus torreyana*), II: 456
— white (*Pinus strobus*), II: 289, 316–317, 635
— whitebark, III: 375
pine beetle infestation, III: 17
pinyon-juniper woodland, II: 303, 500, 504, 632–633; III: 194

plankton diversity, II: 360
planning, II: 11; III: 516–517
— collaborative, III: 268–273
— interdisciplinary, III: 18, 297
— strategic, II: 376, 412
— time-span, II: 338–340
planning horizon, I: 50; II: 345, 404
planning process, I: 244; II: 4; III: 61
planning technology, II: 9
planning unit, ecological, II: 346
plant community, I: 55; II: 55, 364; III: 648
plant cover, II: 55
plant cultivation, II: 448
plant gathering, II: 450
plant growth, II: 444
plant harvest, II: 458
plant husbandry, II: 450
plant indicator, II: 401
plant macrofossil, II: 460
plant productivity, II: 12, 72
plant reproduction, I: 32
plant use, II: 459
plantation, II: 74, 539; III: 332, 351
— Douglas-fir, III: 523
— pine, I: 192; II: 204, 248, 293, 526, 588
Pleistocene, II: 479
plover, western snowy, II: 654
pluralism, I: 136
Poisson process, II: 105, 109
polarization, strategic, III: 144
policy, I: 123, 242; III: 297
— administrative, I: 107
— environmental, III: 609
— fair market value, I: 80
— federal, II: 179
— fire exclusion, I: 195
— governmental, III: 147
— let-it-burn, II: 321
— monetary, III: 389
— natural resource, II: 435
— Northern New Mexico, III: 200–201
— public, I: 80
— restrictive, I: 140
policy change, II: 14; III: 34–36
policy maker, I: 230, 232; III: 539
policy making, II: 412
policy shift, III: 29, 35
policy strategy, III: 671
policy tool, incentive-based, I: 198, 199
politics, II: 3, 10–11, 30; III: 102, 674
pollen analysis, I: 67, 72; II: 225, 322, 325, 475, 477, 485, 607
pollen sediment, I: 98
pollinator, II: 62
pollutant, II: 590 *see also* contaminant
— air, I: 98; II: 51; III: 213, 298, 698
— water, I: 98
pollutant detector, II: 12
pollution, I: 16, 31; II: 6; III: 327, 577
— air, I: 178; III: 124, 334

— industrial, II: 212
— non-point source, II: 640; III: 24
— point source, II: 653, 658
— toxic, III: 167
pollution charge, III: 448
pollution control, III: 11, 24
pollution prevention, III: 167
polygon, II: 402; III: 611, 652
polymorphism, II: 115
ponderosa pine forest partnership (PPFP), II: 575–576
ponderosa pine/bunchgrass woodland, II: 289
ponderosa pine/savanna community, II: 207
population, I: 170; II: 24, 27, 87–134; III: 309
— ceiling, II: 104–105
— local, II: 128
— minimum viable (MVP), II: 88, 104, 148
— viable, II: 561; III: 18, 354, 443
population abundance, II: 93
Population and Habitat Viability Analysis (PHVA), II: 125
population attribute, II: 144
population biology, II: 40
population change, I: 23; II: 95
population collapse, II: 101
population cycle, I: 31; II: 180
population density, II: 104, 183
population dynamics, I: 19, 23; II: 45, 51, 69–70, 92, 96–99, 113, 141
population genetics, II: 92
population growth, II: 77
— rate, I: 170, 172; II: 51, 63, 64–66, 120, 301
population model, II: 69–71, 92–113
population monitoring, II: 124
population size, I: 21, 22, 31; II: 27, 46, 51, 69, 89, 103–104, 107–108, 117, 143, 149, 689
population status, II: 140
population viability analysis (PVA), I: 19–26; II: 40, 42, 58, 66, 89, 92, 117–124, 127–128, 135–156, 334; III: 569
poverty, II: 32; III: 232
prairie, II: 59
— shortgrass, II: 81
prairie dog (Cynomys spp.), II: 164, 180
pre-assessment planning, I: 217
precautionary principle, I: 188; II: 44; III: 14, 408, 445
precipitation, II: 57
precision level, III: 578
predation, I: 15; II: 56, 62, 69–70, 257
predator, II: 52, 73, 75, 77
predator–prey system, II: 69, 101–102, 447, 479, 639
predictability, II: 522
preemption principle, III: 28
Prescott National Forest, II: 636
preservation value, III: 389
prey, II: 52
price, I: 178, 179; III: 350, 390
— regional, I: 180
pricing, I: 188; III: 400
princess tree, II: 304
principal component analysis, II: 68, 361–362
Prisoner's Dilemma, III: 135, 137

private sector, II: 177–179
privatization model, III: 385–387
privet, common, II: 637
probability, I: 192; II: 90; III: 456
problem defining, I: 219
problem solving, III: 101, 145, 249
process
— biological, I: 226; II: 227–228
— definition, II: 220
— deterministic, II: 112
— fluvial, II: 260
— prescriptive, III: 310
— tectonic, II: 219, 224–227
procurement agreement, III: 116
producer, I: 198
product, non-timber, III: 333
product life, I: 80
product market, II: 537
production
— agricultural, II: 523, 686
— algal, II: 64
— export-oriented, II: 679
— maximum sustainable, III: 406
— primary, I: 34, 36; II: 46, 49–51, 66, 74–78, 227–228, 233–234; III: 407 see also net primary production
— secondary, II: 50, 74, 78
production function, II: 222; III: 323–324
productivity, I: 15; II: 40, 49–50, 55, 57, 63–65, 68
— biological, II: 652
productivity gradient, II: 63
productivity loss, II: 656
productivity management, II: 74–77
programming
— dynamic, III: 703
— fuzzy, III: 703–704
— linear, III: 702, 705
— stochastic, III: 703
property
— common, III: 136, 263, 267
— private, III: 263
property right, I: 5, 197–198; II: 5, 29, 32; III: 13, 26, 136, 149, 263, 361, 386, 438, 464–468
property right transfer, III: 399
property tax, I: 198; III: 470
prospect theory, III: 422
protection standard, III: 424
pruning, II: 452; III: 194
psychology, III: 61, 310
ptarmigan, willow, II: 105
Public Affairs Information Service (PAIS), III: 49
public awareness, III: 120, 134
Public Benefit Rating System (PBRS), III: 470
public expectation, I: 115–121; III: 300
public good, III: 387
public interest, III: 60, 69
public involvement, I: 81; II: 541; III: 3, 20–22, 74–75, 78–83, 99, 106, 111, 113, 133, 158, 173, 236, 271, 310, 454, 554, 599, 677, 697
public meeting, III: 46

public participation, I: 85; II: 541; III: 19, 25, 59, 114,
     155–156, 162, 201, 212, 217, 219, 247, 305, 453, 513–514,
     538, 554–555, 588, 599, 697
Pueblo IV period, I: 74; II: 501–502
pulp mill, III: 334, 367
pulse harvesting, III: 349
pyrodendrochronology, II: 460–461

quality assessment, III: 553
quality assurance, I: 92; II: 594
quality control, I: 229; III: 31, 587, 593, 596–597, 607
quality management, III: 341
questionnaire
— standardized, III: 50
— Threats, III: 591
questionnaire return rate, III: 68
questionnaire survey methodology, III: 45

rabbit, II: 75
— cotton-tail (Sylvilagus floridanus), II: 124
rabbit brush, rubber, II: 407
racoon, II: 259
radiocarbon dating, II: 478, 481, 485
railroad, I: 68
rainforest, tropical, II: 684
ranching
— cattle, I: 74
— historic, II: 475
— small-scale, III: 197
random allocation, I: 211; III: 518
random utility (RU) model, III: 391
randomization, II: 15
Range of Variability Approach (RVA), II: 318
rangeland, I: 90–91, 182; II: 567
Rangeland and Forest Resources Planning Act (RPA), I:
     109, 174; III: 337, 536
rangeland restoration, II: 636
rapid application development (RAD), III: 633
Rapid Assessment Procedure (RAP), III: 189, 202–203
Rapid Ethnographic Assessment Project (REAP), III: 203,
     213
Rapid Rural Appraisal (RRA), III: 52, 203, 267
raptor, II: 76
rarity, II: 54, 67, 73, 143, 203, 640
rat, kangaroo, II: 53
rationality
— scientific, III: 284
— technical, I: 161; III: 248–249, 667, 673
rattlesnake, II: 164
reasoning modeling, III: 290
reclamation, I: 96, 100, 101; II: 602
Record of Decision (ROD), III: 513, 646
recovery, II: 42, 77, 297–298, 620
recovery planning, III: 36
recovery zone, II: 149
recreation, I: 180, 182, 189; II: 29–30, 524, 551, 565, 636; III:
     17, 112, 175, 220, 322, 353, 361
recreation activity, III: 368, 371
recreation area, III: 270

recreation impact management, II: 648
recreation industry, II: 26
recreation preference, III: 329
recreation value, III: 394
recreation visit day (RVD), III: 395
recycling, I: 80, 173
red cedar, western (Thuja plicata), II: 451, 607, 649
Red Queen's hypothesis, II: 117
redbud (Cercis occidentalis), II: 450, 452
Redfield ratio, II: 229
reductionism, III: 87, 284, 301
redundancy, II: 40, 240–241, 245, 362
redundant backup system, II: 52
redwood, II: 26
Redwood National Park, III: 15
reed, common (Phragmites australis), II: 610; III: 236
reference condition, I: 72, 73, 95, 98; II: 256, 269, 313, 316,
     319, 404, 416, 471, 476, 496, 573, 604, 607, 624; III: 441
reference range of variability (RRV), II: 414
reference site, II: 416–417
reforestation, II: 176, 529; III: 475
reforestation, urban, III: 264
refuge, I: 87; II: 287
regeneration, I: 194; II: 300
— natural, II: 576; III: 349
regeneration cutting, II: 322
region, I: 36; II: 222; III: 309
regression
— linear, II: 68
— non-linear, II: 68
regulation, II: 11, 135; III: 34, 150, 452
rehabilitation, I: 95, 96, 100, 101; II: 602, 661–662
relatedness, phylogenetic, II: 48
relationship
— cause and effect, II: 323, 397, 522, 594; III: 457–458
— core-periphery, II: 504
— human-environment, III: 299
— intercultural, I: 148
— vegetation-site, II: 382, 406
— wildlife-habitat, II: 332, 361
release assessment, III: 436
reliability analysis, III: 419
religion, II: 29; III: 194
remediation, II: 6, 226, 658, 660
remote sensing, I: 17, 37, 76, 234–235, 238; II: 9, 12, 185–186,
     332, 364, 375–376; III: 570, 610–612, 620, 631, 642, 653,
     655, 657
replication, I: 211; II: 15
reproduction, II: 87–134
reproduction rate, I: 22; II: 69, 106, 111
research, I: 98, 211; III: 509–510
— anthropological, III: 203–204
— ethnobotanical, II: 459–460
— ethnographic, II: 458–459; III: 221
— inductive, I: 117
— paleoenvironmental, II: 460
— qualitative, I: 120
research method, I: 115
research natural area (RNA), II: 202

research objective, I: 116
research program, II: 14
reservation era, III: 10
reserve, II: 533
— core, I: 32
reserve area, II: 201
reserve design, II: 195, 201
reserve system, aquatic, II: 573
resilience, I: 13, 33–35, 90; II: 13, 41, 49, 53, 223, 229, 249, 334, 404, 585–586; III: 349, 405, 437 *see also* ecosystem resilience
— economic, III: 374
resiliency *see* resilience
resistance, I: 34–35; II: 13, 41, 114, 223, 249; III: 287
resolution, I: 47, 49; II: 310, 332, 337, 345
resource
— archaeological, II: 497–502
— cultural, II: 24, 27–28
— natural, II: 24
— non-market, I: 181; II: 11
— non-renewable, II: 569–571
— renewable, I: 79
— sharing, I: 136
— socioeconomic, II: 24, 27
resource advisory council (RAC), II: 567
resource allocation, III: 346–347
resource availability, I: 40; III: 143, 233
resource conservation, I: 80
resource consumption, I: 169, 170, 173; II: 552
resource demand, II: 558
resource economics, I: 185–190; III: 383–412
resource exploitation, II: 257–258, 370, 460, 590; III: 138
resource export, II: 474
resource information technician (RIT), III: 619
resource inventory, I: 57–58; II: 43, 377, 397, 400, 402–403 *see also* inventory
resource law, III: 26, 10
resource management, I: 58, 136; II: 4, 30, 377; III: 209–226, 283, 297
resource output, I: 78–79; II: 538–539
resource planning, I: 215, 218; II: 377, 557
resource procurement, II: 474
resource production, II: 517, 551–552
resource sampling *see* sampling
resource selection function (RSF), II: 122–123
resource use, I: 34, 145–150; III: 692
— seasonal, III: 217
— shift, I: 169–175
resource value, II: 566
Resources Planning Act, III: 348, 536, 542
respiration, I: 34; II: 224, 227–228, 233–234
respiratory loss, II: 50
response
— adaptive, III: 433
— biological, II: 161
— evolutionary, II: 42, 235
responsibility, I: 246; II: 9, 15
— civic, I: 136; III: 133
— ethical, I: 73

— legal, I: 73
responsiveness, I: 162
— civic, I: 155; III: 233
— social, I: 79, 80; II: 519
restoration, I: 73, 96, 99; II: 275, 321, 487–488, 496, 601–602, 605, 659
— estuarine, III: 168–169
— hands-off, III: 524
— hands-on, III: 524
— riparian, II: 261, 640–650; III: 175
— terrestrial, II: 630–640
restoration decision factor, I: 100
restoration design, II: 623–626
restoration ecology, I: 95–104; II: 8, 433, 436, 459, 599–618, 662
restoration forestry, II: 634
restoration management, II: 664
restoration strategy, II: 621
retention era, III: 10
revenue, III: 375, 400, 435
— excess, III: 376
— marginal, III: 378
— net, III: 390–391
reversibility principle, II: 44
reward system, I: 233; III: 527, 609
rhinoceros, II: 79
rice grass, Indian (*Oryzopsis hymenoides*), II: 452
Richard's Center-periphery model, II: 688–689
Rio 'Earth Summit' conference, II: 7, 39, 543
Rio Grande National Forest, II: 575
riparian community, I: 30; II: 641
riparian ecosystem, I: 33, 37, 95–97; II: 255, 489; III: 576
riparian forest, II: 168
riparian function, II: 170–171
riparian quality, III: 437
riparian vegetation, II: 270, 488
riparian zone, I: 30; II: 157, 169–170, 271, 488–489, 624, 644–645; III: 378, 424, 445
ripple effect, III: 349
risk, I: 192, 194; II: 11, 693; III: 415–416, 423, 432–433
— acceptable, I: 192; III: 424, 445
— de minima, III: 426, 445
— regional, III: 423
risk adjustment, III: 433–434, 436, 442, 445–453, 457
risk agent, III: 433, 435 *see also* stressor
risk assessment, I: 100, 191–195, 243; II: 89, 102, 127, 262, 561–563, 628; III: 35, 413, 416–420, 434–444, 457
risk attitude, III: 421–422, 425–426, 455–459
risk aversion, I: 79, 192; III: 324, 421–422, 426, 447
risk behavior, III: 420–423
risk characterization, I: 193; III: 433, 439–444, 457
risk communication, III: 420, 433, 453–454, 457
risk effects paradigm, III: 432
risk endpoint, III: 417, 433, 436, 453
risk estimation, III: 436, 439
risk evaluation, III: 433–434, 444–445, 457
risk factor, I: 20–21; II: 140–141
risk management, I: 192, 193; III: 324, 413, 423–426, 431–462
risk perception, III: 415, 420

risk policy, I: 195; III: 425, 447
risk reduction, I: 79, 193; III: 425
risk seeking, III: 421
risk source, III: 433, 436
risk standard, III: 425
risk trade-off, I: 194; III: 445
risk-taking, II: 15, 22
river system type, II: 366
road, I: 42; II: 567
road building, II: 78, 561; III: 378
road corridor, II: 161–162, 413
road drainage, II: 642
road management, II: 643
roadkill, II: 293
Roadless Area Review and Evaluation (RARE), II: 526
robustness, II: 112, 693
Rocky Mountain National Park, I: 117, III: 61
Rogue National Forest, III: 253
rotation length, II: 74–76, 78, 80,286, 301, 574; III: 397
rust
— blister, III: 375
— fusiform, I: 192

Safe Drinking Water Act, III: 24
Safe Harbor Agreement, I: 199; III: 476–477
sage scrub, II: 78
sagebrush steppe, II: 52, 299, 304, 407
Saguaro National Park, Arizona, III: 175
salamander, II: 198
— cheat mountain (Plethodon nettingi), II: 210
— Red Hill, II: 124
salinity, II: 258, 367, 610
salmon, I: 20, 130, 232; II: 26, 28, 620; III: 125
— Atlantic, II: 410
— Pacific, II: 270; III: 440
— West Coast (Oncorhynchus spp.), II: 167
salmon fishery, III: 607
sample, I: 228
sample size, III: 48, 578–579, 586
sampling, I: 57, 229, 237; II: 412
— gradient-oriented, III: 556
— grid-based, II: 67
— non-statistical, I: 238; III: 645
— point, III: 610–611
— random, II: 67; III: 46, 54, 70, 586, 594, 612, 656
— snowball, III: 51
— statistical, I: 238; III: 645
— systematic, II: 67; III: 594
sampling design, I: 227; II: 67–68; III: 557, 578–579, 586–587, 593–594
sampling intensity, III: 578
sampling method, I: 117, 237, 238; II: 61; III: 543, 596, 645
sampling strategy, III: 69
San Bernardino National Forest, III: 143
San Diego, II: 213–214
San Isabel National Forest, III: 451
San Juan National Forest, Colorado, II: 575; III: 213
sanctioning, III: 138
sand habitat, III: 580

Santa Fe National Forest, New Mexico, I: 150; II: 489; III: 199, 212
satellite imagery, I: 17, 55, 238; II: 57, 66–67, 71, 199, 206, 207, 211, 366; III: 146, 630, 647
satellite population, II: 91
savanna, II: 79
sawmill, I: 68; III: 367
scale, I: 33, 47–53, 73, 151, 153, 177–183, 225, 233; II: 41–43, 157, 331–352; III: 637
— analytical, II: 347
— broad, II: 333
— coarse, II: 333
— ecoregional, II: 203
— extent, II: 295, 332, 345
— fine, II: 333
— geographic, II: 128; III: 632
— global, I: 36; II: 222; III: 310
— grain, II: 226, 295, 345
— large, II: 345
— size, II: 332
— small, II: 333, 345
— spatial, I: 29, 34, 49, 50, 69, 212, 216, 220; II: 9, 159–160, 197–198, 263, 294, 331, 343, 347–348, 475; III: 13–14, 537, 540–542, 608, 664
— temporal, I: 29, 34, 49, 50, 69, 212, 216; II: 160–161, 197–198, 222, 294, 331, 343, 476; III: 13–14, 537, 541, 608, 664
scale-change effect, II: 159
scarcity, III: 322
— economic, III: 347
— of knowledge, III: 508
scenario, succession, II: 59
scenario analysis, II: 404, 411, 591, 630, 640, 650, 656
scenario development, III: 426, 435, 543
scenario planning, I: 220; III: 419, 558–559
science, II: 9–10
— civic, II: 11
— experimental, I: 132
Scientific Panel on Late Successional Ecosystems, II: 10
sculpin, deepwater, II: 410
sea lamprey, II: 410
sea otter, II: 52
sea urchin, II: 52, 590
sedge (Carex spp.), II: 453
sedimentation, II: 172, 220, 226–227
seed monies, III: 143
seed tree system, II: 634–635
self-certification, II: 544–545
self-learning, III: 302
self-organization, II: 248, 315
self-regulation, III: 283
self-restoration, II: 13
sense of place, III: 138, 140, 157, 163, 539
sense of threat, III: 163
sensitivity analysis, II: 96, 113, 413; III: 440, 680, 702
sequoia, giant, I: 40; II: 301, 454
serpentine soil, I: 17; II: 166
set aside, II: 550; III: 350–352, 445, 476, 480
Severeid's Rule, III: 314

sewage, II: 75
sewage plant, II: 234
shading, II: 589, 647
Shannon–Weiner Index, II: 56
Shawnee National Forest, Illinois, II: 631
sheep, bighorn (*Ovis canadensis*), II: 447, 644; III: 194, 511
shelterwood, II: 634–635
Sherman Anti-Trust Act, I: 140; III: 147
Shorebank Corporation, III: 473
Shorebank Enterprise Pacific, III: 473
ShoreBank Pacific, III: 473
shoreline marsh, II: 382
shortgrass prairie, II: 444
shortleaf pine-bluestem system, I: 210; II: 322, 630; III: 528
shovel testing, I: 74–75; II: 507
Sierra Nevada Ecosystem Project (SNEP), I: 216, 217; III: 536, 538, 541
signal to noise ratio, I: 92; II: 594
significance indicator, II: 508
significance level, III: 578
silviculture, II: 273, 320, 464, 578, 609, 634; III: 281, 364
Simpson's Index, II: 56
simulation model, I: 90, 243–244; II: 106, 144–145; III: 290, 438, 510, 567–571, 615, 691, 706
sink, II: 158, 653
sink area, II: 63
Siskiyou National Forest, III: 253
site, III: 309
— archaeological, I: 75, 98; III: 218
— industrial, I: 100
— sacred, I: 148
— salient, I: 73
site classification, II: 369, 412
site effects, II: 338
site index, II: 72, 74
site management, III: 61
site productivity, I: 208
Siuslaw National Forest, II: 572; III: 517
skeleton weed, rush (*Chondrilla juncea*), II: 181
snag, I: 194; II: 80, 221, 282; III: 450
snowberry (*Symphoricarpus racemosus*), II: 452
social behavior, III: 285
social capital, I: 164
social ecology, III: 186, 279, 285–289
social impact assessment (SIA), II: 32
social order, II: 24–25, 31–32
social science, I: 115, 118, 119; III: 3, 44–45, 48, 55, 60, 245, 283
Social Science Index (SSI), III: 49
social theory, I: 159–161; II: 23; III: 245–278
social welfare, I: 189; II: 13; III: 385–386, 389
social well-being, III: 361, 541
society, dynamics of, I: 163
sociology, I: 154, 159–165; II: 23; III: 227, 310
— environmental, II: 23
software, III: 703
— ALEX, I: 23
— application, I: 243
— limitations, II: 337

— RAMAS, I: 23
— VORTEX, I: 23
soil, I: 55; II: 49
Soil and Water Resources Conservation Act, II: 525
soil contamination, II: 76
soil erosion, II: 76, 32, 531
soil fertility, II: 51, 57, 230
soil formation, I: 31; II: 60
soil process, II: 606
soil productivity, I: 91; II: 64; III: 545
soil retention, III: 397
soil survey, II: 420
solar constant, II: 223
sourberry (*Rhus trilobata*), II: 450
source, II: 158
source area, II: 63
source-sink model, II: 108
Southern Appalachian Assessment (SAA), III: 62, 66, 538–540
sowing, II: 452
space, ecological, II: 237–242
sparrow, Bachman's, II: 588
Spatial Unified Data Dictionary (SPUDD), III: 651
Special Emphasis Program Manager (SEPM), III: 216
speciation, II: 58
species
— early-successional, II: 303
— endangered, I: 109, 224; II: 54, 66, 73, 109, 125, 135, 258, 455, 529, 628, 644; III: 13, 30, 176, 334, 354, 477, 605, 625
— endemic, II: 27, 54, 58, 61, 151, 357; III: 269
— exotic, II: 40, 45, 47–48, 52–53, 60, 75–76, 78, 80, 101, 181–182, 188–189, 205, 292, 304, 410, 453, 488, 530, 563, 631, 637, 656; III: 334
— functional group, I: 15
— invasive, I: 16, 21, 44; II: 54–56, 60, 81, 181–182, 206 *see also* exotic species
— native, II: 47–48, 563, 599
— non-game, II: 79
— non-native *see* exotic species
— sensitive, II: 204; III: 626
— shade-tolerant, II: 50, 59
— territorial, II: 100
species abundance *see* abundance
species analysis, II: 362
species area curve, II: 61–62, 67
species assessment, II: 150–152
species composition, II: 325
species distribution, I: 55; II: 137
species diversity, I: 13–18, 81; II: 39–40, 58, 66–72; III: 625
  see also biodiversity and diversity
species extinction *see* extinction
species loss, II: 317
species management, II: 80–81, 138
species movement, II: 161
species persistence *see* persistence
species reintroduction, II: 60, 640, 648; III: 590
species replacement, II: 60
species response, II: 286
species richness, I: 55; II: 27, 55, 65, 121, 162–163, 170, 240, 303, 602; III: 545

species survey, II: 67
spider flower, stemmed, II: 654
spider, Wolf, II: 161
spillover, I: 187; III: 387
sport utility vehicle (SUV), III: 335
Spring Mountains National Recreation Area (SMNRA), I: 160–161; III: 268–273
spruce
— black, III: 238
— red (*Picea rubens*), II: 210, 316
— Sitka (*Picea sitchensis*), II: 335
— white, III: 238, 281
spurge, leafy (*Euphorbia esula*), I: 16, 22, 31, 99; II: 181, 637–638
squirrel, ground, II: 53
stability, II: 693; III: 285, 437
stakeholder, I: 48, 87, 91, 126, 135, 148, 177, 181, 185, 188, 189, 202, 216, 218; II: 11, 15, 411, 571, 590–592; III: 6, 13, 23, 62, 89–91, 97, 106, 158, 195, 210, 259, 289, 323, 360–362, 662
stakeholder analysis, III: 266
stakeholder identification, III: 106, 116, 119–120, 140, 213, 366, 368, 377
stakeholder involvement, III: 109
stakeholder objective, III: 694
stakeholder participation, III: 177
stand, I: 36; II: 222, 332, 608; III: 309
stand density, II: 635
stand scale, II: 265
stand-replacement dynamics, II: 321
standard, III: 639–640
— command-and-control, III: 444, 446
— compatibility, III: 17
— regulatory, I: 192
— safe minimum, III: 426, 445
— zero-risk, III: 426
standardization, II: 398; III: 249
standard forest mensuration method, II: 67
starthistle, yellow (*Centaurea solstitialis*), II: 181, 638
State Historic Preservation Officer (SHPO), III: 218
Statistical Analysis System (SAS), III: 66
Statistical Procedure for the Social Sciences (SPSS), III: 66
Steens Mountain, Oregon, I: 117
Stewardship Incentive Program (SIP), III: 481
stochasticity
— demographic, I: 20; II: 61, 102–103, 107, 114, 338
— environmental, II: 102, 107, 114, 338
— sources, II: 90
strategic thinking, III: 105
strategy, III: 302
— cooperative, III: 137
stratification, social, I: 154
stratigraphic position, II: 481
stream
— pristine, I: 17
— reference site, II: 416
stream channel, II: 169, 271, 641; III: 581, 595
stream erosion, II: 172
stream quality, II: 32

stream reach, III: 595
stream restoration, II: 643; III: 521
stress, I: 36, 87, 90; II: 224, 242, 585, 629
stressor, I: 36; II: 27, 88, 224, 235, 246, 410; III: 432–433 *see also* risk agent
structure, I: 99; II: 621
— age, II: 93–95
— genetic, II: 116–117
— geologic, II: 165
— physical, I: 15
— spatial, II: 45, 56–58, 90–91, 114–115
— stage, II: 93–95
— trophic, II: 49–50, 65–66, 585
structure indicator, II: 265–266
Structured Query Language (SQL), III: 614
subregion, III: 161
subsidy, I: 35–36, 200; II: 535; III: 400, 448
subsistence, III: 222
subsistence pattern, II: 500
subsistence use, III: 233, 237, 242, 330
subsistence value, III: 77
succession, I: 15, 33, 34, 43; II: 51, 58–60, 70, 219, 228, 233–237, 242, 256, 282, 291–292, 298, 301, 366, 651
— allogenic, II: 59
— autogenic, II: 59
— cyclic, II: 236
— ecological, II: 359
— primary, II: 59
— secondary, II: 59, 301
succession model, II: 22
succession pattern, II: 242
sumac (*Rhus trilobata*), II: 452
Sumter National Forest, South Carolina, II: 324
surface flow, II: 652
Surface Mining Control and Reclamation Act, I: 602; III: 24
surface remain, I: 74, 75
surrogate, II: 118, 142–143, 355, 360
surrogate measure, II: 124, 592
survey, I: 74–75; III: 50, 59, 235, 701
— archaeological, II: 507
— cross-sectional, III: 68
— door-to-door, III: 50
— household, I: 118; III: 67–71
— influencing factors, I: 119
— longitudinal, III: 68
— mail, I: 119; III: 50, 68, 70
— on-site, III: 62, 71–74
— public, III: 79–80
— telephone, III: 50, 62, 70
— visual counter, III: 596
survey design, III: 67
survey research, I: 118, 120; III: 49–50
survey response rate, III: 46
survival, II: 62–63
survival rate, II: 70, 96, 102–103, 111, 649
sustainability, I: 1, 17, 79, 85, 89, 228; II: 7–8, 13, 25, 33, 258, 435–436, 517, 521, 523, 569, 583; III: 12, 89, 299, 366, 407, 437 *see also* ecosystem sustainability
sustainability indicator, II: 587

Sustainable Forestry Initiative (SFI), III: 472
sustenance, II: 26, 30, 33
switch grass (*Panicum virgatum*), II: 610
symbiosis, II: 101, 231; III: 203
synergism, II: 288
synthesis approach, II: 370–371
system
— integrated, I: 241
— knowledge-based, I: 243–244
— open, I: 90
— organizational, III: 664
— sociocultural, I: 126; II: 25
— sociotechnical, II: 503–504
— taxonomic, II: 420
system dynamics, II: 358
system state, III: 568
system transition, III: 568
system variable, III: 568
system, economic, I: 126, 178
systems approach, I: 179; III: 287, 346, 663, 671
systems limits, II: 358–359
systems theory, I: 242; II: 27; III: 300, 510, 663
systems thinking, I: 126, 131; II: 14

Tahoe Coalition of Recreation Provider (TCORP), III: 75
tallgrass prairie, I: 17, 41, 96; II: 444, 528, 537; III: 406, 464
tallgrass–midgrass–shortgrass prairie, II: 630
tamarisk, II: 52, 649
Tangled Bank hypothesis, II: 117
tax, I: 197; III: 360, 400, 483
tax benefit, III: 466
tax credit, III: 448, 470
tax incentive, II: 213; III: 448
tax policy, I: 179, 198; III: 464–465, 468
tax reduction, III: 466
tax revenue, earmarked, III: 482
taxonomy, II: 46, 396; III: 229
— cladistic, I: 153
— Linnean, I: 153
taxpayer, II: 691
Taylor Grazing Act, III: 11, 26
technological capital, I: 186
technology, I: 80, 169; II: 677, 691; III: 334–335
— new, I: 68, 234
technology change, III: 365
technology sharing, III: 159
technology transfer, I: 161, 209; III: 249–250, 258
temperature, II: 57, 68
terminology, standardized, III: 606
The Nature Conservancy (TNC), I: 23, 56, 58, 147; II: 41, 67, 203, 364, 368, 422–423; III: 169, 173, 235, 270, 334, 467, 479, 648
Theodor Roosevelt National Park, I: 99
thermocline development, II: 408
thermodynamics, I: 34; II: 65
— first law of, II: 223; III: 401
— second law of, II: 223; III: 401
thermoluminescence, II: 481

thinning, II: 79–80, 242, 300, 464, 486, 572, 578–579, 605–606, 608–609, 635; III: 375, 509
thistle
— musk, II: 638
— Russian, II: 52
— Scotch, II: 638
threshold, II: 91, 100–101, 182; III: 525, 564, 572
— artifact-density, I: 75
threshold, quantitative, III: 579
thrush, wood (*Hylocichla mustelina*), II: 334
tidal cycle, II: 610
tiger, II: 79
tilling, II: 452–453
timber, I: 178, 187
timber certification, I: 199; III: 468
timber economy, I: 182
timber growth rate, III: 332
timber harvest, I: 86, 179, 244; II: 270, 464, 521, 524, 545, 575, 606; III: 18, 331, 362, 378
timber import, II: 694, 696
timber industry, III: 144, 162
timber price, II: 697
timber product, II: 695–696
timber production, I: 78
timber removal, II: 486
timber sale, II: 29, 535, 537, 562–563, 578; III: 163, 374
Timber Sale Program Analysis System (TSPAS), III: 375
Timber Sale Program Information Reporting System (TSPIRS), III: 375
timber shortage, I: 180
timber value, III: 364
timber volume, standing, II: 530
toadflax, dalmatian, II: 638
Toiyabe National Forest, III: 269
tolerance range, I: 73
Tongass National Forest, Alaska, II: 270, 570
tortoise
— desert, II: 113, 124
— gopher, II: 589
tourism, II: 213; III: 175, 333, 351
tourist industry, II: 29
Toxic Substances Control Act (TSCA), II: 525
toxicity, II: 166–167, 657, 660
toxin, II: 75
tradable development permit, III: 448, 477
trade, I: 67
— global, I: 180; III: 366
— international, III: 349
trade-off, I: 78, 133, 177, 179, 180, 186, 189, 234; II: 11, 126, 696, 700–701; III: 323, 347, 359, 362–364, 370, 413
— identification, III: 378
— risk–benefit, III: 425, 446
— risk–cost, III: 424
traditional ecological knowledge (TEK), I: 156; III: 237–239
trajectory
— stable, II: 296
— successional, II: 233
transformation theory, III: 251
transition zone, II: 376

transplanting, II: 453
travel cost (TC), III: 391, 394–398, 407
Treaty of Guadalupe Hidalgo, III: 200
tree growth, II: 32, 478
tree life cycle, II: 563
tree mortality, I: 31, 40; II: 70, 207, 235, 289, 321, 454, 604;
    III: 334
tree planting, II: 649
tree ring dating, I: 67, 72, 98; II: 460, 478, 506, 575, 594
tree vole, red (*Aborimus longicaudus*), II: 608
triangulation, III: 48, 54
trophic level, II: 22, 71, 228, 367, 410, 585
trout
— Apache (*Oncorhynchus apache*), III: 520
— blue ribbon, I: 180
— bull, II: 629; III: 33
— cutthroat, II: 629
— Lahonton cutthroat, II: 644
— lake, II: 410
— rainbow, II: 585
— redband, II: 646
turkey, II: 26
— wild, II: 622; III: 334
turnip, prairie (*Psoralea esculenta*), II: 450
turtle, green sea, II: 113
type
— ecological, II: 420
— functional, II: 46–47, 60
typology, III: 229

U.S. Army Corps of Engineers, I: 58; II: 6
U.S. Department of Agriculture (USDA), III: 466
U.S. Environmental Protection Agency (EPA), I: 58, 193
U.S. Fish and Wildlife Service (USFWS), I: 58, 110; II: 424;
    III: 23–24, 142, 619
U.S. Forest Service (USFS), I: 109, 130, 182; II: 5, 398, 424;
    III: 168, 236
U.S. Geological Survey (USGS), I: 58; II: 186, 382, 423, 488,
    559–660; III: 169, 642, 648
U.S. Government Accounting Office (USGAO), II: 626, 645
uncertainty, I: 191–195, 244; II: 5, 8, 11, 16, 40, 90, 114, 589;
    III: 299, 324, 335, 388, 414–416, 432–433, 451, 559, 709
uncertainty, regulatory, III: 452
uncertainty, scientific, II: 259; III: 14, 20, 35, 98
uncertainty analysis, III: 419
uncertainty assessment, III: 418, 436
understanding, I: 242–243; III: 453
— biological, III: 566
— conceptual, I: 225
— cyclical II: 449
— ecological, III: 133, 570
— environmental, III: 149, 211
— linear structures, II: 449
— of stakeholder, III: 5, 47
— scientific, II: 548
— sociocultural, III: 45, 216–217, 357
Unfunded Mandates Reform Act, III: 25, 178
United Nations Development Programme (UNDP), III: 338

United Nations Educational, Scientific and Cultural
    Organization (UNESCO), II:
urban ecology, II: 214
urban forestry, III: 475
Urban Greening Initiative (UGI), III: 219
Urban Resource Initiative (URI), III: 265–267
urbanization, I: 31; II: 681–682, 695; III: 335
user pays principle, III: 354

validation monitoring, III: 31, 589–590
valuation, I: 185, 189
— contingent, II: 592
— economic, III: 363, 384
— individualistic, I: 189
— monetary, I: 186
value, I: 136, 177; II: 28; III: 59, 261, 403, 406–408
— aesthetic, II: 54, 699
— American, III: 132
— commercial, III: 394
— cultural, I: 5, 173, 189; II: 698
— direct use, I: 186; III: 389
— ecological, II: 9; III: 308, 390
— economic, II: 628; III: 77, 363
— environmental, III: 51, 308
— ethical, III: 242
— individualistic, I: 186
— instrumental, III: 389–390
— monetization, I: 189
— non-instrumental, III: 389–390
— non-market, II: 592; III: 359
— non-quantifiable, I: 189
— non-use, III: 387
— public use, III: 308
— scenic, III: 11
— shifting, I: 115–121; II: 551; III: 59–84, 100
— social, I: 131, 187; II: 3, 6–7, 13, 628; III: 89, 213, 308
— spiritual, III: 308
variability, II: 693; III: 287, 415
— range, II: 317–319, 478, 496
— natural, I: 92, 98; III: 572
— spatial, II: 56
variation
— genetic, II: 115–117
— historic, II: 472
vegetation, I: 353
— existing, II: 363
— historic, II: 534
— old growth, II: 60
— potential natural (PNV), I: 28; II: 209, 356, 363, 369, 371,
    422
vegetation classification, II: 363
vegetation cover, II: 403, 605
vegetation degradation, II: 489
vegetation dynamics, II: 573
vegetation management, II: 268
vegetation map, II: 337, 356
vegetation mosaic, II: 630
vegetation pattern, II: 506

vegetation physiognomy, II: 356
vegetation recovery, II: 78
vegetation structure, II: 408
vegetation type, II: 55
vegetation unit, II: 363
vertebrate, II: 123
viability analysis, III: 376, 442
viability assessment, II: 137
videography, I: 238
volatilization, II: 232
volcanism, II: 224–225
vole (*Microtus* spp.), II: 338
— red-backed, II: 293
von Humboldt, I: 53
voluntary initiative, I: 201; III: 150, 464–465, 484–486
Voyageurs National Park, III: 592

wallflower, Menzie's (*Erysimum menziesii*), II: 111
warbler, yellow (*Dendroica petechia*), II: 337, 340
Washington Forest Practices Board (WFPB), II: 574
waste management, III: 141
waste water, II: 173; III: 114–115
water, II: 26, 68
water availability, II: 653
water budget, II: 229
water chemistry, II: 57
water conservation, II: 6
water consumption, II: 173
water flow, I: 35; II: 314
water purification, I: 201
water quality, I: 178; II: 54, 64, 275, 530, 647, 653; III: 30, 66, 70, 168, 176–178, 592
water resource, II: 26
water right, II: 31–32; III: 35, 195, 237
water supply, II: 73; III: 176
water temperature, II: 644, 647
water uptake, II: 593
waterfowl, I: 201; II: 75–76, 382, 654–655; III: 194, 464
watershed, I: 35–36; II: 158, 222, 343, 353, 547; III: 580
watershed analysis, I: 37–38, 236; II: 212, 255, 261, 269–273, 626
watershed attributes, II: 271
watershed level, III: 647
watershed management, II: 4, 646
watershed model, II: 324, 366
watershed morphology, II: 272
watershed protection, I: 17; III: 24
watershed restoration, II: 633
weathering, II: 165, 226, 275
weed control technique, II: 637
weed infestation, II: 638
weed management tool, II: 488
Weibull model, II: 296, 298
welfare improvement, III: 389–392
welfare indicator, III: 405
Western Cascades, Pacific Northwest, II: 264–266
Western Rangeland Riparian Management, I: 37; II: 260–261

wet meadow riparian community, I: 37
wetland, I: 187; II: 531, 650–656
— artificial, II: 661
— riparian, II: 65, 566
wetland classification, II: 367, 382, 398, 402
wetland degradation, II: 651
wetland function, II: 651, 655
wetland reserve program, I: 200; III: 481
wetland system, I: 96
wheatgrass, bluebunch (*Agropyron spicatum*), II: 53
White Mountain National Forest, III: 366, 395
wide area network (WAN), III: 650
Wild and Scenic Rivers Act, I: 108; II: 6, 525; III: 11
wild dog, African (*Lycaon pictus*), II: 117
wild rice (*Zizania aquatica*), II: 452–453
wild rye, Salina, II: 407
wilderness, I: 86, 178; III: 16–17, 197
Wilderness Act, I: 108–109; II: 6, 525–526, 569, 622; III: 11, 16, 197
wildfire, I: 87, 124, 193; II: 53, 268, 434 *see also* fire
wildlife, II: 26
wildlife corridor, II: 293; III: 175
wildlife habitat, I: 244
wildlife habitat model, II: 123–124, 142
wildlife preservation, III: 15, 23
wildlife production, III: 353
wildlife refuge, III: 11, 17–18
Willamette National Forest, Oregon, II: 320, 572–574
willingness, I: 132, 162
willingness-to-accept, I: 189; III: 391, 403
willingness-to-change, II: 628
willingness-to-compromise, II: 73
willingness-to-learn, III: 157, 529
willingness-to-listen, III: 179
willingness-to-pay, I: 186, 189; II: 592; III: 370, 386, 389, 391, 395–396, 403, 407
willow, III: 511
willow (*Salix* spp.), II: 337, 447, 453, 644, 649
willow community, I: 37
windthrow, I: 41; II: 78, 291
wiregrass (*Aristida* spp.), II: 286, 315, 588
wise use movement, III: 27
wolf (*Canis lupus*), I: 99, 100; II: 12, 447, 592, 621, 640
— dire (*Canis dirus*), II: 446
— Mexican grey, III: 472
— reintroduction, I: 199
wood consumption, II: 524, 700
wood product, demand, III: 331
wood production, I: 82
wood substitution, II: 521; III: 351
wood use efficiency, III: 335
woodcat, dusky-footed (*Neotoma fuscipes*), II: 455
woodpecker
— pilated (*Dryocopus pileatus*), II: 562
— red-cockaded (*Picoides borealis*), I: 28, 210; II: 80, 209, 257, 316, 322, 455, 620, 639–640; III: 477, 528
woody debris, II: 80, 221, 270, 272, 319, 562
World Bank, III: 203, 338
World Conservation Union (IUCN), I: 22; II: 357; III: 334

world economy, II: 503
world landscape, II: 505–507
World Resources Institute (WRI), I: 174; III: 338
world trade network, I: 67; II: 474
world view, I: 131; II: 28; III: 261, 567
World Wide Web (WWW), II: 12, 186; III: 339-340, 658,
    697–699
World Wildlife Fund (WWF), II: 544
Worldwatch Institute, I: 174; III: 338
worst case assessment, III: 419, 439

yaupon (*Ilex vomitoria*), II: 453

Yellowstone National Park, I: 42, 100, 109; II: 77, 119, 211,
    293, 296, 447, 454–455, 648, 694; III: 11, 475, 511
yew tree, Pacific, III: 475
yield curve, II: 71
Yosemite National Park, II: 211
Youth Conservation Corps (YCC), III: 219
Yukon Delta National Wildlife Refuge, III: 619

zone system, biogeoclimatic, II: 370
zoogeography, II: 357
zooplankton, II: 585
Zurich–Montpellier Tradition, II: 364

List of Abbreviations

ACEC	Area of Critical Environmental Concern
ACLD	Augusta Creek landscape design
AGRICOLA	National Agricultural Library Catalogue
AHCS	Aquatic Habitat Classification System
AHP	Analytical Hierarchy Process
AIHP	American Industrial Heritage Project
AIRFA	American Indian Religious Freedom Act
ARPA	Archaeological Resource Protection Act
AMA	Adaptive Management Area
AMS	Accelerator Mass Spectrometry
ANILCA	Alaska National Interest Lands Conservation Act
ANOVA	Analysis of Variance
ANSI	American National Standards Institute
APAEA	Asian Pacific American Employee Association
ARC	Angeles Restoration Crew
ARPA	Advanced Research Projects Agency
ARS	Agricultural Research Service
ASTM	American Society for Testing and Materials
ATLSS	Across Trophic Level System Simulation
ATV	All Terrain Vehicle
AUM	Animal Unit Month
AVHRR	Advanced Very High Resolution Radiometer
AVIMS	Annual Vegetation Inventory and Monitoring System
BART	Best Available Retrofit Technology
BBN	Bayesian Belief Network
BBS	Breeding Bird Survey
BIA	Bureau of Indian Affairs
BLM	Bureau of Land Management
BMP	Best Management Practice
BOR	Bureau of Reclamation
BPR	Business Process Re-engineering
BRD	Biological Resource Division
CA	Conjoint Analysis
CAB	Commonwealth Agricultural Bureau
CanFI	Canadian National Forest Inventory
CASE	Computer-Aided Systems Engineering
CBC	California Biodiversity Council
CBSG	Captive Breeding Specialists Group

CEN	Committee for European Normalisation
CEP	California Environmental Project
CEQ	Council on Environmental Quality
CERCLA	Comprehensive Environmental Response, Compensation and Liability Act
CERES	California Environmental Resources Evaluation System
CFIC	Canadian Forest Inventory Committee
CODA	Conservation Options and Decisions Analysis
COTS	Commercial Off The Shelf
CPR	Common Pool Resource
CRM	Coordinated Resource Management
CSDS	Common Survey Data Structure
CUSTOMER	Customer Use and Survey Techniques for Operations, Management, Evaluation and Research
CV	Contingent Valuation
CWHR	California Wildlife Habitat Relationship
CWMA	Cooperative Weed Management Area
DBH	Diameter at Breast Height
DEM	Digital Elevation Model
DEMO	Demonstration of Ecosystem Management Options
DER	Department of Environmental Regulation
DFC	Desired Future Condition
DNR	Department of Natural Resources
DOI	Department of the Interior
DOQ	Digital Orthophoto Quadrangles
DSS	Decision Support System
EA	Environmental Assessment
ECOMAP	Ecological Classification and Mapping
ECS	Ecological Classification System
EIS	Environmental Impact Statement
ELF	Extremely Low Frequency
ELT	Ecological Land Type
ELTP	Ecological Land Type Phase
EM	Ecosystem Management
EMDS	Ecosystem Management Decision Support
ENR	Everglades Nutrient Removal
EPA	Environmental Protection Agency
ERU	Ecological Reporting Unit

ESA	Endangered Species Act	LSF	Late Serial Forest
ESA	Ecological Society of America	LTA	Land-Type Association
ESRI	Environmental Systems Research Institute	LTER	Long Term Ecological Research
FACA	Federal Advisory Committee Act	LUHNA	Land Use History of North America
FAPRI	Food and Agriculture Policy Research Institute	LUMP	Land Unsuitable Mining Program
		MA	Management Area
FEMAT	Forest Ecosystem Management and Assessment Team	MANOVA	Multivariate Analysis of Variance
		MLRA	Major Land Resource Area
FERC	Federal Energy Regulatory Commission	MOA	Memorandum of Agreement
FGDC	Federal Geographic Data Committee	MOU	Memorandum of Understanding
FIA	Forest Inventory and Analysis	MRLC	Multi-Resolution Land Characterization
FIFRA	Federal Insecticide, Fungicide, and Rodenticide Act	MSS	Landsat Multi-Spectral Scanner
		MUSY	Multiple-Use Sustained Yield Act
FIP	Forest Incentive Program	MVP	Minimum Viable Population
FIPS	Federal Information Processing Standard	NAGPRA	Native American Graves Protection Act
FLPMA	Federal Land Policy Management Act	NALC	North American Landscape Characterization
FOIA	Freedom of Information Act		
FS	Forest Service	NAPAP	National Acid Precipitation Assessment Program
FSC	Forest Stewardship Council		
FWS	Fish and Wildlife Service	NAPP	National Aerial Photography Program
GAO	Government Accounting Office	NASA	National Aeronautics and Space Administration
GAP	Gap Analysis Program		
GDP	Gross Domestic Product	NAWQA	National Water Quality Assessment
GDSS	Group Decision Support System	NBII	National Biological Information Infrastructure
GIS	Geographic Information System		
GLO	Government Land Office	NBS	National Biological Service
GPP	Gross Primary Productivity	NED	Northeast Decision Model
GPS	Global Positioning System	NEP	Net Ecosystem Productivity
HACU	Hispanic Association of Colleges and Universities	NEPA	National Environmental Policy Act
		NFMA	National Forest Management Act
HCP	Habitat Conservation Plan	NFS	National Forest System
HEP	Habitat Evaluation Procedure	NGO	Non-Governmental Organization
HSI	Habitat Suitability Index	NHFEU	National Hierarchical Framework for Ecological Units
HSI	Hispanic Serving Institution		
HTM	Habitat Transaction Method	NHPA	National Historic Preservation Act
ICBEMP	Interior Columbia Basin Ecosystem Management Project	NIPF	Nonindustrial Private Forest
		NIST	National Institute for Standards and Technology
ICPSR	Inter-University Consortium for Political and Social Research		
		NMFS	National Marine Fisheries Service
IDH	Intermediate Disturbance Hypothesis	NOAA	National Oceanographic and Atmospheric Administration
IDT	Interdisciplinary Team		
IE	Information Engineering	NPP	Net Primary Production
IEC	International Electrotechnical Commission	NPS	National Park Service
IEM	Integrated Environmental Management	NPV	Net Present Value
INA	Information Needs Assessment	NRCS	Natural Resources Conservation Service
IPCC	Intergovernmental Panel on Climate Change	NSDI	National Spatial Data Infrastructure
		NSF	National Science Foundation
IRI	Integrated Resource Inventory	NSFHWAR	National Survey of Fishing, Hunting and Wildlife Associated Recreation
IRS	Internal Revenue Service		
ISC	Interagency Scientific Committee	NSRE	National Survey on Recreation and the Environment
ISG	Interagency Steering Group		
ISO	International Standards Organization	NTMB	Neotropical Migratory Birds
IUCN	International Union for Conservation of Nature	NWFP	Northwest Forest Plan
		OGC	Office of General Council
LACC	Los Angeles Conservation Corps	OLA	Opportunity Los Angeles
LAN	Local Area Network	OMB	Office of Management and Budget
LCM	Life-Cycle Management	OPA	Oil Pollution Act
LRMP	Land and Resource Management Plan	PAIS	Public Affairs Information Service
LRR	Land Resource Region	PAR	Participatory Action Research

PBRS	Public Benefit Rating System
PFC	Paper Functioning Condition
PHVA	Population and Habitat Viability Analysis
PIF	Partners in Flight
PNC	Potential Natural Community
PNV	Potential Natural Vegetation
PPFP	Ponderosa Pine Forest Partnership
PRA	Participatory Rural Appraisal
PVA	Population Viability Analysis
RAC	Resource Advisory Council
RAD	Rapid Application Development
RAP	Rapid Assessment Procedure
RARE	Roadless Area Review and Evaluation
REAP	Rapid Ethnographic Assessment Project
RIT	Resource Information Technician
RNA	Research Natural Area
ROD	Record of Decision
RPA	Rangeland and Forest Resources Planning Act
RRA	Rapid Rural Appraisal
RRV	Reference Range of Variability
RSF	Resource Selection Function
RU	Random Utility
RVA	Range of Variability Approach
SAA	Southern Appalachian Assessment
SAS	Statistical Analysis System
SEPM	Special Emphasis Program Manager
SFI	Sustainable Forestry Initiative
SIA	Social Impact Assessment
SIP	Stewardship Incentive Program
SHPO	State Historic Preservation Officer
SMNRA	Spring Mountains National Recreation Area
SNEP	Sierra Nevada Ecosystem Project
SPSS	Statistical Procedure for the Social Sciences
SPUDD	Spatial Unified Data Dictionary
SQL	Structured Query Language
SSI	Social Science Index
SUV	sport utility vehicle
TC	Travel Cost
TCORP	Tahoe Coalition of Recreation Provider
TCR	Total Cost of Risk
TEK	Traditional Ecological Knowledge
TM	Landsat Thematic Mapper
TNC	The Nature Conservancy
TSCA	Toxic Substances Control Act
TSPAS	Timber Sale Program Analysis System
TSPIRS	Timber Sale Program Information Reporting System
UGI	Urban Greening Initiative
UNDP	United Nations Development Programme
UNESCO	United Nations Educational, Scientific and Cultural Organization
URI	Urban Resource Initiative
USDA	U.S. Department of Agriculture
USFS	U.S. Forest Service
USFWS	U.S. Fish and Wildlife Service
USGAO	U.S. Government Accounting Office
USGS	U.S. Geological Survey
WAN	Wide Area Network
WFPB	Washington Forest Practices Board
WRI	World Resource Institute
WWF	World Wildlife Fund
WWW	World Wide Web
YCC	Youth Conservation Corps

Contents of Volumes I–III

◆ VOLUME I: KEY FINDINGS

Foreword vii
Preface ix
Acknowledgments xv
Introduction—The Ecological Stewardship Project: A Vision for Sustainable Resource Management 1

BIOLOGICAL AND ECOLOGICAL DIMENSIONS
Genetic and Species Diversity 13
Population Viability Analysis 19
Ecosystem and Landscape Diversity 27
Ecosystem Processes and Functioning 33
Disturbance and Temporal Dynamics 39
Scale Phenomena 47
Ecological Classification. 53

HUMANS AS AGENTS OF ECOLOGICAL CHANGE
Human Roles in the Evolution of North American Ecosystems 63
Cultural Heritage Management 71
Producing and Using Natural Resources 77
Ecosystem Sustainability and Condition 85
Ecological Restoration. 95

PUBLIC EXPECTATIONS, VALUES, AND LAW
Legal Perspectives 107
Public Expectations and Shifting Values 115
The Evolution of Public Agency Beliefs and Behavior. 123
Processes For Collaboration. 129
Regional Cooperation 135

SOCIAL AND CULTURAL DIMENSIONS
Cultural/Social Diversity and Resource Use 145
Social Classification 151
Social Processes. 159

ECONOMIC DIMENSIONS
Demographics and Shifting Land and Resource Use 169
Economic Interactions at Local, Regional, and National Scales. 177
Ecological and Resource Economics 185
Uncertainty and Risk Assessment 191
Economic Tools For Ecological Stewardship 197

INFORMATION AND DATA MANAGEMENT

Adaptive Management 207
Assessment Methods. 215
Monitoring and Evaluation 223
Data Collection, Management, and Inventory 231
Decision Support Systems. 241

Index 247
List of Abbreviations 277
Contents of Volumes I–III 281

◆ VOLUME II

Foreword vii
Preface xi
Acknowledgments xvii
Introduction—The Ecological Stewardship Project: A Vision for Sustainable Resource Management xxi

INTRODUCTION

Ecosystem Management: Evolving Model for Stewardship of the Nation's Natural Resources 3
   — *Hanna J. Cortner, John C. Gordon, Paul G. Risser, Dennis E. Teeguarden, and Jack Ward Thomas*
The Human Ecosystem as an Organizing Concept in Ecosystem Management. 21
   — *Gary E. Machlis, Jo Ellen Force, and William R. Burch, Jr.*

BIOLOGICAL AND ECOLOGICAL DIMENSIONS

Biological and Ecological Dimensions — Overview. 39
   — *Gary K. Meffe*
A Functional Approach to Ecosystem Management: Implications for Species Diversity 45
   — *Michael Huston, Gary McVicker and Jennifer Nielsen*
Population Viability Analysis: A Primer on its Principal Technical Concepts 87
   — *Barry R. Noon, Roland H. Lamberson, Mark S. Boyce, and Larry L. Irwin*
Population Viability in Ecosystem Management. 135
   — *Richard S. Holthausen, Martin G. Raphael, Fred B. Samson, Dan Ebert, Ronald Hiebert, and Keith Menasco*
An Ecosystem Approach for Understanding Landscape Diversity 157
   — *James R. Gosz, Jerry Asher, Barbara Holder, Richard Knight, Robert Naiman, Gary Raines, Peter Stine, and T.B. Wigley*
Describing Landscape Diversity: A Fundamental Tool for Ecosystem Management 195
   — *Julie A. Concannon, Craig L. Shafer, Robert L. DeVelice, Ray M. Sauvajot, Susan L. Boudreau, Thomas E. DeMeo, and James Dryden*
Ecosystem Processes and Functioning. 219
   — *Ariel E. Lugo, Jill S. Baron, Thomas P. Frost, Terrance W. Cundy, and Phillip Dittberner*
Ecosystem Processes and Functions: Management Considerations 255
   — *Steven J. Paustian, Miles Hemstrom, John G. Dennis, Phillip Dittberner, and Martha H. Brookes*
Disturbance and Temporal Dynamics. 281
   — *Peter S. White, Jonathan Harrod, William H. Romme, and Julio Betancourt*
Practical Applications of Disturbance Ecology to Natural Resource Management. 313
   — *R. Todd Engstrom, Sam Gilbert, Malcolm Hunter, David Merriwether, Greg Nowacki, and Page Spencer*
Scale Considerations for Ecosystem Management. 331
   — *Jonathan B. Haufler, Thomas Crow, and David Wilcove*
Scales and Ecosystem Analysis. 343
   — *David L. Caraher, Arthur C. Zack, and Albert R. Stage*
Principles for Ecological Classification. 353
   — *Dennis H. Grossman, Patrick Bourgeron, Wolf-Dieter N. Busch, David Cleland, William Platts, G. Carleton Ray, C. Richard Robins, and Gary Roloff*

The Use of Ecological Classification in Management 395
  – *Constance A. Carpenter, Wolf-Dieter N. Busch, David T. Cleland, Juan Gallegos, Rick Harris, Ray Holm, Chris Topik, and Al Williamson*

HUMANS AS AGENTS OF ECOLOGICAL CHANGE
Humans as Agents of Ecological Change — Overview 433
  – *Nels C. Johnson*
Native American Influences on the Development of Forest Ecosystems 439
  – *Thomas M. Bonnicksen, M. Kat Anderson, Henry T. Lewis, Charles E. Kay, and Ruthann Knudson*
Human Influences on the Development of North American Landscapes: Applications for Ecosystem
Management 471
  – *Richard D. Periman, Connie Reid, Matthew K. Zweifel, Gary McVicker, and Dan Huff*
Historical Science, Heritage Resources, and Ecosystem Management. 493
  – *Alan P. Sullivan, III, Joseph A. Tainter, and Donald L. Hardesty*
The Historical Foundation and the Evolving Context for Natural Resource Management on Federal
Lands 517
  – *Douglas W. MacCleery and Dennis C. Le Master*
Ecosystem Management and the Use of Natural Resources 557
  – *Marlin Johnson, James Barbour, David W. Green, Susan Willits, Michael Znerold, James D. Bliss, Sie Ling Chiang, and Dale Toweill*
Ecosystem Sustainability and Condition 583
  – *Ron Carroll, Jayne Belnap, Bob Breckenridge, and Gary Meffe*
Ecological Restoration and Management: Scientific Principles and Concepts 599
  – *Wallace Covington, William A. Nearing, Ed Starkey, and Joan Walker*
Ecosystem Restoration: A Manager's Perspective 619
  – *James G. Kenna, Gilpin R. Robinson, Jr., Bill Pell, Michael A. Thompson, and Joe McNeel*
Human Population Growth and Tradeoffs in Land Use 677
  – *Joel E. Cohen*

Index 703
List of Abbreviations 733
Contents of Volumes I–III 737

◆ VOLUME III

Foreword vii
Preface xi
Acknowledgments xvii
Introduction—The Ecological Stewardship Project: A Vision for Sustainable Resource Management xxi

PUBLIC EXPECTATIONS, VALUES, AND LAW
Public Expectations, Values, and Law — Overview 3
  – *D. Dean Bibles*
Legal Perspectives on Ecosystem Management: Legitimizing a New Federal Land
Management Policy. 9
  – *Robert B. Keiter, Ted Boling, and Louise Milkman*
Understanding People in the Landscape: Social Research Applications for Ecological
Stewardship 43
  – *John C. Bliss*
Shifting Values and Public Expectations: Management Perspectives. 59
  – *H. Ken Cordell, Howie Thompson, Barbara McDonald, Clyde Thompson, Christina Ramos, Gerald Helton, Stephen E. Ragone, and Michelle Dawson*
The Evolution of Public Agency Beliefs and Behavior Toward Ecosystem-Based
Stewardship 85
  – *James J. Kennedy and Michael P. Dombeck*

Collaborative Processes For Improving Land Stewardship and Sustainability 97
  – *Mark Hummel and Bruce Freet*
Regional Cooperation: A Strategy For Achieving Ecological Stewardship 131
  – *Steven L. Yaffee*
Management Perspectives on Regional Cooperation 155
  – *Kathleen M. Johnson, Al Abee, Gerry Alcock, David Behler, Brien Culhane, Ken Holtje, Don Howlett,*
    George Martinez, and Kathleen Picarelli

SOCIAL AND CULTURAL DIMENSIONS
Social and Cultural Dimensions — Overview 183
  – *Margaret A. Shannon*
Cultural and Social Diversity and Resource Use 189
  – *William deBuys, Muriel Crespi, Susan H. Lees, Denise Meridith, and Ted Strong*
Resource Management Strategies For Working With Cultural and Social Diversity 209
  – *Carol Raish, Lynn Engdahl, William Anderson, Donald Carpenter, Muriel Crespi, Paul Johnson, Les*
    McConnell, and Earl Neller
Social Classification in Ecosystem Management 227
  – *Maureen H. McDonough, Don Callaway, L. Marie Magelby, and William Burch, Jr.*
Some Contributions of Social Theory to Ecosystem Management 245
  – *J. Kathy Parker, Victoria E. Sturtevant, Margaret A. Shannon, William R. Burch, Jr., J. Morgan Grove,*
    Jeremiah C. Ingersoll, and Lois Sagel
Ecosystem Management — Some Social and Operational Guidelines for Practitioners 279
  – *William R. Burch, Jr. and J. Morgan Grove*
Toward an Ecological Approach: Integrating Social, Economic, Cultural, Biological, and
Physical Considerations 297
  – *Roger N. Clark, George H. Stankey, Perry J. Brown, James A. Burchfield, Richard W. Haynes,*
    and Steven F. McCool

ECONOMIC DIMENSIONS
Economics and Ecological Stewardship — Overview 321
  – *Robert Mendelsohn*
Shifting Human Use and Expected Demands on Natural Resources 327
  – *Steven K. Cinnamon, Nels C. Johnson, Greg Super, Judy Nelson, and David Loomis*
Economic Interactions at Local, Regional, National, and International Scales 345
  – *Roger A. Sedjo, Dale E. Toweill, and John E. Wagner*
Understanding Economic Interactions at Local, Regional, National and International Scales 359
  – *Amy L. Horne, George Peterson, Kenneth Skog, and Fred Stewart*
Ecological and Resource Economics as Ecosystem Management Tools 383
  – *Stephen Farber and Dennis Bradley*
Uncertainty, Risk, and Ecosystem Management 413
  – *Richard Haynes and Dave Cleaves*
Risk Management For Ecological Stewardship 431
  – *David A. Cleaves and Richard W. Haynes*
Tools For Ecological Stewardship 463
  – *Paige Brown*

INFORMATION AND DATA MANAGEMENT
Information and Data Management — Overview 499
  – *Francisco Dallmeier*
Adaptive Management 505
  – *Bernard T. Bormann, Jon R. Martin, Frederic H. Wagner, Gene W. Wood, James Alegria, Patrick G.*
    Cunningham, Martha H. Brookes, Paul Friesema, Joy Berg, and John R. Henshaw
Assessments for Ecological Stewardship 535
  – *Russell T. Graham, Theresa B. Jain, Richard L. Haynes, Jim Sanders, and David L. Cleaves*

Understanding and Managing the Assessment Process. 551
– *Gene Lessard, Scott Archer, John Probst, and Sandra Clark*
Evaluating Management Success: Using Ecological Models to Ask the Right Monitoring
Questions 563
– *David Maddox, Karen Poiani, and Robert Unnasch*
Managing the Monitoring and Evaluation Process 585
– *Timothy Tolle, Douglas S. Powell, Robert Breckenridge, Leslie Cone, Ray Keller, Jeff Kershner, Kathleen*
  S. Smith, Gregory J. White, and Gary L. Williams
Data Collection, Management, and Inventory 603
– *Allen Cooperrider, Lawrence Fox III, Ronald Garrett, and Tom Hobbs*
Information Management For Ecological Stewardship 629
– *Cynthia S. Correll, Catherine A. Askren, Rob Holmes, Henry M. Lachowski, Gust C. Panos, and W. Brad*
  Smith
Decision Support Systems/Models and Analyses 661
– *Chadwick D. Oliver and Mark J. Twery*
Decision Support For Ecosystem Management 687
– *Keith Reynolds, Jennifer Bjork, Rachel Riemann Hershey, Dan Schmoldt, John Payne, Susan King, Lee*
  DeCola, Mark Twery, and Pat Cunningham

Index 723
List of Abbeviations 753
Contents of Volumes I–III 757